WEYERHAEUSER ENVIRONMENTAL CLASSICS

WILLIAM CRONON, EDITOR

Weyerhaeuser Environmental Classics are reprinted editions of key works that explore human relationships with natural environments in all their variety and complexity. Drawn from many disciplines, they examine how natural systems affect human communities, how people affect the environments of which they are a part, and how different cultural conceptions of nature powerfully shape our sense of the world around us. These are books about the environment that continue to offer profound insights about the human place in nature.

The Great Columbia Plain: A Historical Geography, 1805–1910
by D. W. Meinig

Mountain Gloom and Mountain Glory: The Development of the Aesthetics of the Infinite by Marjorie Hope Nicolson

Tutira: The Story of a New Zealand Sheep Station by Herbert Guthrie-Smith

A Symbol of Wilderness: Echo Park and the American Conservation Movement
by Mark Harvey

Man and Nature: Or, Physical Geography as Modified by Human Action
by George Perkins Marsh; edited and annotated by David Lowenthal

Conservation in the Progressive Era: Classic Texts edited by David Stradling

DDT, Silent Spring, *and the Rise of Environmentalism: Classic Texts*
edited by Thomas R. Dunlap

Weyerhaeuser Environmental Classics is a subseries within Weyerhaeuser Environmental Books, under the general editorship of William Cronon. A complete listing of the series appears at the end of this book.

DDT, *SILENT SPRING*, AND THE RISE OF ENVIRONMENTALISM

CLASSIC TEXTS

EDITED BY THOMAS R. DUNLAP

FOREWORD BY WILLIAM CRONON

UNIVERSITY OF WASHINGTON PRESS

SEATTLE AND LONDON

DDT, *Silent Spring*, and the Rise of Environmentalism: Classic Texts *is published with the assistance of a grant from the Weyerhaeuser Environmental Books Endowment, established by the Weyerhaeuser Company Foundation, members of the Weyerhaeuser family, and Janet and Jack Creighton.*

Printed in United States of America
Designed by Pamela Canell
14 13 12 11 10 09 5 4 3 2 1

University of Washington Press, PO Box 50096, Seattle, WA 98145
www.washington.edu/uwpress

Library of Congress Cataloging-in-Publication Data
DDT, Silent Spring, and the rise of environmentalism : classic texts /
edited by Thomas R. Dunlap ; foreword by William Cronon.
p. cm. — (Weyerhaeuser environmental classics)
Includes bibliographical references and index.
ISBN 978-0-295-98834-4 (pbk. : alk. paper)
1. DDT (Insecticide)—Environmental aspects—History—20th century.
2. Environmentalism—History—20th century.
3. Nature—Effects of human beings on—History—20th century.
4. Carson, Rachel, 1907–1964. Silent spring. I. Dunlap, Thomas R., 1943–
TD196.P38D45 2009 363.17′92—dc22 2008020044

The paper used in this publication is acid-free and 90 percent recycled from at least 50 percent post-consumer waste. It meets the minimum requirements of American National Standard for Information Sciences—Permanence of Paper for Printed Library Materials, ANSI Z39.48–1984. ♾ ♻

Contents

PART 2 DDT'S BRIGHT PROMISE AND NEGLECTED PROBLEMS (1942–1958)

DDT as Miracle Chemical

Early Warnings

PART 3 RISING CONCERN ABOUT NEW PROBLEMS

DDT, Food Chains, and Wildlife

PART 4 THE STORM OVER *SILENT SPRING*

PART 5 DDT AND MALARIA

Foreword

SILENT SPRING AND THE BIRTH OF MODERN ENVIRONMENTALISM

WILLIAM CRONON

It may be an oversimplification to say that the modern environmental movement began with Rachel Carson's *Silent Spring*, but it is hard to overstate that book's impact. Prior to its publication in 1962, various environmental concerns were becoming more prominent in the years following World War II. Air and water pollution, radioactive fallout, toxic exposures, overpopulation, loss of wild lands, resource scarcities, suburban sprawl: all of these were gaining new public attention by the early 1960s, and all would become part of the new movement over the course of the decade. But *Silent Spring* was a lightning rod like no other. Gaining visibility first as a series of articles in the *New Yorker* and then as a best-selling book, it catapulted its author—whose earlier writings about the sea had been enormously popular but not at all controversial—into a political firestorm. Carson's indictment of DDT and other insecticides brought heated rebuttals from chemical companies that manufactured these products, from government agencies that promoted their use, and from scientists who believed that their benefits far outweighed their harms. Supporting her were doctors concerned about the medical and genetic consequences of toxic exposure, biologists who feared the impact of pesticides on wildlife populations, and parents worried about possible harm to their children. Carson was called before a Senate committee to testify about the book, and President John F. Kennedy was asked about it by the press. CBS News produced a widely viewed television documentary based on Carson's findings, and newspapers and magazines across the country used the controversy to educate the public about previously unnoticed threats. By the time of the first Earth Day in 1970, *Silent Spring* was universally regarded as one of the founding texts of environmentalism as we know it in the first decade of the twenty-first century.

To understand the movement, then, one needs to understand Carson's book and the controversy it provoked. *Silent Spring* is still required reading for this reason, and the book remains remarkably compelling half a century after it first appeared. But because the world we now inhabit has been so powerfully shaped by Carson's perspectives, it takes a real act of historical imagination to go back to the time before those perspectives were as common as they are today. Until 1962, Americans were far more inclined to regard DDT as a miracle than as a menace. It had protected crops the world over from insect damage amounting to hundreds of millions of dollars, assuring increased food supplies to people who might otherwise have starved. It seemed to be well on the way to eradicating malaria from many parts of the world, saving millions of lives in the process. That it had achieved so much good with so little apparent impact on human health or prosperity seemed nothing less than wondrous.

To appreciate just how widespread and persuasive such views were before *Silent Spring* was published, one must begin by examining documents from much earlier in the century, when DDT was being rediscovered as a powerful new tool—far more effective than its highly toxic predecessors—for attacking some of the oldest insect adversaries of humanity. Thomas R. Dunlap's *DDT,* Silent Spring, *and the Rise of Environmentalism: Classic Texts* makes it much easier for readers to do this, enabling them to trace the unexpected transformation of DDT from environmental hero to villain. Dunlap is an ideal person to compile this collection. A leading environmental historian and historian of science, he authored the first serious scholarly study of the insecticide controversy just a few years after DDT was banned by the U.S. Environmental Protection Agency in 1972. From his deep knowledge of environmental thought in the decades leading up to and following the publication of *Silent Spring,* Dunlap has selected a handful of documents that bring vividly to life this key transformation in American attitudes toward technology and the natural world.

Beginning with examples drawn from the early years of scientific entomology, Dunlap illustrates the hope that selectively toxic chemicals might provide new weapons in "the war on pests"—a common military metaphor that runs through a number of these documents. The earliest such chemicals used in agriculture tended to rely on heavy metals such as lead and arsenic, which were dangerous to insects, mammals, and people alike. Given the obvious threats associated with such compounds, it becomes much easier to understand the exuberance with which DDT was greeted when its highly selective acute toxicity for insects was discovered in 1939. Here, it seemed, was a compound that devastated pests while leaving mammals and other vertebrates virtually unscathed. DDT famously contributed to Allied war efforts during World War II by fighting typhus in the Mediterranean theatre and malaria in the tropics. Excitement about these successes created widespread enthusiasm for potential civilian applications when the war was over, nicely illustrated in Dunlap's collection by "Insect-O-Blitz" advertisements and a popular magazine article asking "How Magic Is DDT"? As DDT began to be used in ever greater quantities, evidence soon began to appear suggesting that it might have unexpected effects on wildlife populations, generating documents that are all the more fascinating because of their close juxtaposition with documents promoting the benefits of the new pesticide. Troubling evidence accumulated across the 1950s, gradually providing the analytical foundation on which Carson would construct *Silent Spring*. Dunlap supplies striking examples of the heated rhetoric that swirled around the book and its author in the months following its publication and ends with documents suggesting that the debate over DDT has not entirely ended. The questions Carson asked in her book remain every bit as important and contested today as they were when she first posed them.

DDT, Silent Spring, *and the Rise of Environmentalism* is a volume in Weyerhaeuser Environmental Books' Classic Texts series, which offers modern readers brief collections of carefully selected, well-annotated documents designed to shed light on important turning points in the history of environmental thought and politics. Although we hope these books will be useful in a wide variety of classroom settings, they are not meant to be simply textbooks. We've recruited scholars with a deep knowledge of the subjects these volumes explore, and have asked them to select documents of uncommon richness, arranged in ways that show ideas unfolding over time while also setting up unexpected juxtapositions that reveal subtleties and complexities that no single text could illustrate on its own. When combined with an extended introductory essay and chapter summaries that alert readers to important themes and ideas, these Classic Texts offer to any interested reader accessible, readable guides to some of the most important debates in the history of environmentalism. There have been few more influential controversies than the one launched by Rachel Carson in 1962, and this brief volume does a fine job of guiding readers through its complexities.

Preface and Acknowledgments

When I got out of the Army and went back to graduate school—this time in history—I asked my adviser about a topic for an MA thesis. He said that with my background (a BA and a couple of years of graduate training in chemistry), a history of DDT would be a natural. I was a bit dismayed, for I was trying to get away from science, but I burrowed into the library and quickly became intrigued. In 1945 DDT had been the modern miracle, harmless to humans and deadly to insects, able to stop epidemics in their tracks, the greatest thing since the atomic bomb, and evidence of the wonders of technology and the progress of Western civilization. In 1970, as I was starting my research, DDT was the modern horror, deadly to many forms of life, the worst thing since the atomic bomb, and evidence of the evils of technology and the decline of Western civilization.

What did this say about Americans' views of science and nature and technology? I decided to find out. With the American ban, declared while I was still doing research, the story seemed about over, and in fact, one lawyer I called to ask about his role in a DDT hearing boomed at me over the phone, "What do you want to talk about DDT for? That's history!" I explained that I was writing my dissertation in history. DDT, though, turned out to be as persistent socially as it was biologically. We still have some DDT in our body fat and we still discuss the chemical. I recently heard an academic refer to calls for interior spraying of DDT to prevent malaria in tones suggesting a Final Solution for Africans or at least African wildlife, while the Internet has many sites blaming Rachel Carson for the deaths of millions or, at the very least, for indifference to their fates.

Because DDT involved big issues like technology, progress, and whether or not we ought to (or could) conquer nature as well as very personal ones—will this stuff give me cancer?—people made all sorts of claims. DDT either disrupted hormonal balances in the body and caused cancer or it was so safe we could eat it (which a few of its defenders did at public talks); it killed birds by disrupting their reproduction or it had no impact on wildlife populations; and banning it was a step forward for civilization or the beginning of a retreat to the caves. An entire book could be devoted to Rachel Carson, portrayed as everything from a fearless crusader to an hysterical old maid. About the only thing everyone agreed on was that she wrote very well. The documents collected here cannot even sample these claims and counterclaims. Instead, they address a single, central issue: how growing knowledge about DDT's effects on wildlife shaped Americans' attitudes toward it and then affected policy. Discussions of pesticide residues on fruit, earthworms and robins, and malaria control all come back to that.

A few words on the editing: Some of the shorter documents appear in their entirety, but most are included as excerpts, and the cutting has been severe for some of the scientific articles. Cuts within sentences have been held to a minimum. Footnotes, if any, have been eliminated, but some references in the text have been retained to indicate the authors' reliance on other work. Full annotation for all of the essays can be found in the original publications. Spelling and punctuation are retained unless they seemed particularly jarring, in which case they were silently corrected. Individuals mentioned in passing are usually not identified. Where it seemed useful for continuity, note has been made of omitted material.

The range of views people have taken suggests that we know nothing for certain about DDT, but that is wrong. We know a lot. Scientists agree that when introduced into the environment, it persists, spreads through air and water, and concentrates up food chains—that is where the DDT in our body fat comes from. At the quantities used in the postwar years it affected reproduction in some species of wildlife, particularly birds at the top of food chains. Scientists saw that happen in the wild and confirmed it by testing eggs, measuring eggshells going back to 1900, and feeding DDT to captive birds. There seems no reason to believe that DDT levels in the environment harm human health. We can reasonably be concerned about something we all have in our bodies but do not know is safe and a little queasy about breast-fed babies getting DDT from their mother's milk, but as a human health hazard the worst we can say is the old verdict allowed in Scotch law: "Not proven." We do not *need* it for crop production or, in most cases, for public health. It is useful in tropical countries and in some instances, if not necessary, it is the best thing to use. Finally, seeing the issue in terms of "DDT" or "no DDT" is far too simple.

These readings place the debates about the chemical in the larger perspective of ideas about nature in our industrial society by showing the gradual discovery of its environmental impact and people's reactions to that knowledge. With a few exceptions, they come from the United States. A German chemist first synthesized the compound, the Swiss showed it would kill insects, and its residues drifted around the world, but American manufacturers made it a high-volume, low-cost pesticide, American scientists did most of the early research on its environmental effects, the American public led the charge against it, and American action shaped legislation and practice around the world (strange as that last phrase sounds). The articles come from scientific journals, magazines, newspapers, and government reports, and they range in tone from the passionate to the scientifically exact. All of them, though, bear on what people thought about the relation of humans to the world and the knowledge they had to support their views. I chose them because they spoke to particular parts of that developing argument, but I do not pretend to trace all the connections among them. That will be left, as the mathematics textbooks used to say, as an exercise for the reader.

Bill Cronon and Lita Tarver suggested this project and cheered it on—without them it never would have come into being—but I have other debts as well. Pete Cannon, my old roommate from graduate school, used the expertise gained in a career in Wisconsin's Legislative Reference Bureau to track down some things; the Wisconsin State Historical Society sent me a copy of the tape I made in July 1973, when I interviewed ornithologist Joseph J. Hickey; and behind every permission notice lurks an editor or staffer and sometimes a line of people who helped track down an author or copyright holder. Judy Mattson in the history department at Texas A&M scanned in all the documents for me, making editing much easier. Finally, Texas A&M and its history department provided a stimulating and pleasant place to work. I thank one and all.

DDT, *SILENT SPRING*, AND THE RISE OF ENVIRONMENTALISM

Introduction

From the start of mass production in World War II to the American ban in the early 1970s, DDT's heyday lasted barely thirty years, but its story begins well before that. The ideas and attitudes that guided its use and underlie the continuing debate about its use go far back in Western civilization and American history. From the seventeenth century, science's increasing success in explaining how the world worked gave people confidence that they could understand nature, and nineteenth-century technologies like steam engines, railroads, telegraphs, machine guns, and electric lights encouraged them to believe they could conquer it. Americans believed as well that nature formed an important part of the nation and that contact with nature gave people health and insight. So they celebrated the conquest of the frontier while they lobbied for national parks and made John Muir a nature saint. They saw science as the handmaid of Progress (with a capital P) but wanted it to guide them to an encounter with nature (or Nature) that would lead to insight into Ultimate Reality.

In the late nineteenth century the early chemical pesticides (to lapse into period prose) marched shoulder to shoulder in the vanguard of civilization with electric lights and telephones, but everyone knew they were dangerous. Farmers sprayed apples with lead arsenate, orange growers fumigated their trees with hydrogen cyanide, and home gardeners sprinkled Paris green (copper aceto-arsenite) on their cabbages. The apple sprays caused the most alarm, for they went on as sticky solutions that clung for days or weeks, and people, picking up fruit from a bin in the grocery store months later and thousands of miles away, had no idea what had been applied or washed off. Worse, lead and arsenic accumulated in the body, and small doses, each harmless, could build up and cause sickness or death. When the Bureau of Chemistry began checking for residues in the 1920s, part of its work enforcing the Pure Food Act, it naturally used medical definitions of toxicity and worried most about people who worked around the sprays or ate lots of fresh produce, because lead and arsenic had been known since antiquity to be deadly.

In the late 1930s Swiss chemists began testing synthetic organic chemicals as part of a search for new pesticides, and in 1942 the Geigy company sent samples of one, trade-named Gesarol, to its sales agents in America. The agents promptly turned them over to the War Food Administration, which found that the active ingredient, which killed insects with astounding efficiency and did not seem to harm people, was a chemical with the formidable chemical name of 1,1,1 trichloro, 2, 2*bis* (parachlorophenyl) ethane. That name never became popular and it soon became known as DDT, from the initial letters of its short chemical name: dichloro diphenyl trichloroethane. The government quickly ordered it into production, for in every European conflict typhus killed more people than combat, and the center of the Pacific war, the South Pacific islands and East Asia, harbored a wide array of insect-borne diseases. By 1944 American factories were churning out a million pounds a month, and military public health teams dusted millions of soldiers and civilians with 5 percent DDT powder or fogged their living spaces with aerosols, while planes blanketed islands so the mosquitoes would be dead when the Marines landed. In 1945 production reached 36 million pounds a year.

Its effectiveness, along with postwar optimism about technology raised by wartime miracles like DDT and penicillin, made its use seem natural, even inevitable. These were the years when popular magazines (perfect background reading for the first two sections of this book) told us that in the year 2000 we would have rocket cars, helicopters in every garage, inexpensive bubble houses, and regular rockets to the moon. Houses would clean themselves, meals could be cooked in a few minutes, and television would educate all the children (maybe those last two did come true). Farmers used DDT on fruits, vegetables, cow barns, and dairy cows; towns sent spray trucks down tree-lined streets to get rid of mosquitoes; housewives used the new "bug bombs" in their kitchens; and before the Salk vaccine became available in 1954 a few desperate communities even sprayed schoolchildren to ward off polio. Public health programs around the world relied on DDT. It cost little, posed no significant risk of accidental poisoning (it took a heaping handful of the pure chemical to stand a good chance of killing you), and replaced labor-intensive methods like drying up puddles and pools in abandoned tires and expensive ones like window screens and bed nets. Production peaked in the late 1950s at 180 million pounds a year, enough to give every man, woman, and child in the country their own one-pound bag. If DDT had, like the arsenates, stayed where people put it or had quickly broken down into harmless chemicals all would have been well, but it did not. In the late 1940s the Food and Drug Administration found that DDT sprayed in cow barns showed up in the milk, even when the cows never came in contact with the spray. By 1950 samples from across the country showed that virtually all Americans had some DDT in their body fat.

The public paid little attention to the slow discovery of DDT's unexpected properties, but other news made it fearful of technology. By the mid-1950s people knew that rain carried radioactive fallout from atomic and hydrogen bomb tests and that the radioactive isotope strontium-90 fell on the grass the cows ate and so appeared in milk, and from there went with its chemical cousin calcium into children's bones when they drank their quart a day for strong bones and healthy teeth. That raised the specter of leukemia, the Great Parent Fear of those years. The cranberry scare of 1959 showed just how sensitive the public had become by the end of the decade. Just before Thanksgiving the FDA announced it had found traces of a pesticide that caused cancer in rats in some shipments of cranberries. Sales fell dramatically and some restaurants stopped serving it. Since some 90 percent of the fruit went for sauce for Thanksgiving dinner, public concern caused panic to the offices of Ocean Spray. The company issued press releases, and executives ate handfuls of berries in front of press cameras to show they were safe.

It was the mass spraying campaigns of the late 1950s that focused public attention on pesticides. Towns across the Northeast and upper Midwest used DDT to kill the beetles that carried Dutch elm disease, and homeowners found dead and dying robins on their lawns and fewer songbirds in the trees. In 1959 aerial sprays on Long Island to halt gypsy moths, which were spreading out of New England, produced sticky coatings on cars and homes, complaints, and a lawsuit. The USDA's attempts to kill fire ants on millions of acres across the South by aerial dropping of granules dosed with another chemical, heptachlor, brought vocal opposition from farmers, nature lovers, biologists, and hunters, for it killed fish and birds. Wildlife scientists did more than protest. They applied for research grants, and studies of dead robins on lawns led to new knowledge of DDT's effects. From the late 1950s to the late 1960s a disturbing picture emerged. DDT, introduced into the environment, persisted and concentrated up food chains, with deadly results to species at the top.

Rachel Carson followed each step of this developing story, first as a Fish and Wildlife Service scientist and editor, then as a full-time nature writer. A debate in the letters column of the *Boston Herald* in 1958 pushed her to get more information for a planned article on pesticide problems, and, already suffering from the cancer that would eventually kill her, she turned that into a book, *Silent Spring*.[1] Its publication became an event when, just as the *New Yorker* magazine printed advance excerpts, Americans learned that thalidomide, an apparently safe medicine, had caused the birth of several thousand deformed babies in Europe. Some had no arms, just hands attached to their shoulders; others lacked legs. The National Agricultural Chemicals Association, the pesticide manufacturers' trade group, increased interest by sending speakers around the country who not only disputed Carson's conclusions but charged that she was opposed to science, Progress, and the highest ideals and greatest accomplishments of Western civilization—our conquest of nature. *Silent Spring* appeared after a noisy summer, and it caused an even louder fall and winter.

Circumstances made the book a sensation, but Carson's case made it an inspiration. Like earlier muckraking books, including Arthur Kallett and F. J. Schlink's *100,000,000 Guinea Pigs*[2] (to which it was compared), it warned of dangers to humans from chemicals, but Carson spoke of problems deadlier and less obvious than those posed by additives in food, drugs, and cosmetics:

> Along with the possibility of the extinction of mankind by nuclear war, the central problem of our age has become the contamination of man's total environment with such substances of incredible potential for harm—substances that accumulate in the tissues of plants and animals and even penetrate the germ cells

1. Linda Lear, *Rachel Carson* (New York: Henry Holt, 1997), 312–16.

2. Arthur Kallett and F. J. Schlink, *100,000,000 Guinea Pigs* (New York: Grossett and Dunlap, 1933).

> to shatter or alter the very material of heredity upon which the shape of the future depends. (p. 8)

> [We have put] poisonous and biologically potent chemicals indiscriminately into the hands of persons largely or wholly ignorant of their potential for harm and allowed them to be used with little or no advance investigation of their effects on soil, water, wildlife, and man himself. Future generations are unlikely to condone our lack of prudent concern for the integrity of the natural world that supports all life. (p. 13)

Silent Spring set out a possible future, presented the evidence for disaster, and ended with ways to avoid it. The bulk of the book described the course of pesticide residues through air, water, and soil, their effect on plants, fish, and birds, the spraying campaigns against the gypsy moth, Dutch elm disease, and the fire ant, and the evidence for unknown, long-term effects on humans. That part roused controversy but her recommendations and judgments much more, for she believed that in the long run chemicals failed because insects developed resistance, and the search for newer and more powerful ones only led farther down a road that was "deceptively easy, a smooth superhighway on which we progress with great speed, but at its end lies disaster. The other fork of the road . . . offers our last, our only chance to reach a destination that assures the preservation of our earth" (p. 277). The way out lay not in banning chemicals or abandoning insect control but in using new kinds of controls, some in use now, others being tested, still others only yet ideas, but all "based on an understanding of the living organism they seek to control, and of the whole fabric of life to which these organisms belong" (p. 278).

> As crude a weapon as the cave man's club, the chemical barrage has been hurled against the fabric of life—a fabric on the one hand delicate and destructible, on the other miraculously tough and resilient, and capable of striking back in unexpected ways. . . . The control of nature is a phrase conceived in arrogance, born of the Neanderthal age of biology and philosophy, when it was supposed that nature exists for the convenience of man. . . . It is our alarming misfortune that so primitive a science has armed itself with the most modern and terrible weapons, and that in turning them against the insects it has also turned them against the earth. (p. 297)

The reasons for disagreement over *Silent Spring* ran much deeper than Carson's evidence. Economic entomologists felt the book undermined their authority and denigrated their work. When the University of Wisconsin radio station, WHA, read *Silent Spring* on "Chapter a Day," the Department of Entomology produced a nineteen-page reply arguing that pesticide regulation should be "put back into the hands of the professionals in the College of Agriculture. . . . We agricultural scientists have been given responsibility for making pest control recommendations."[3] Many speakers who challenged Carson's interpretation of scientific papers directed their strongest fire against her belief that we should live with nature and work with its processes. Acceptance of her ideas, one critic said, meant "the end of all human progress, reversion to a passive social state devoid of technology, scientific medicine, agriculture, sanitation. It means disease, epidemics, starvation, misery, and suffering."[4]

The public debate over *Silent Spring* quickly turned into a campaign against DDT, which became the chemical people loved to hate. In 1968 the then-new Environmental Defense Fund challenged the use of DDT in Wisconsin, but even before the Department of Natural Resources' hearing examiner gave his verdict in the summer of 1969, Michigan had banned it and regulated other pesticides as dangers to the environment (and to the commercial and sport fishing so important to the state's economy). In 1972, after extensive public hearings, the Environmental Protection Agency (EPA), established just two years before, ended all but public health uses of DDT in this country.

Even at a time when environmental action had the sanctity of motherhood and apple pie, the ban involved not doing "the right thing" but balancing different "good" things. The EPA hearing examiner, Edward Sweeney, found that by medical definitions DDT did not injure humans and that the harm to birds could not be shown to come from permitted uses. The agency's head, William Ruckelshaus, took a different approach. With regard to people, he judged it better to be safe than sorry. Everyone had DDT in their body fat, and we should know if it was safe or not. As for wildlife, while permitted uses of DDT could not be shown to be the source

3. Critique of 'Insecticides and People,'" 17. Unpublished and privately circulated critique of a panel discussion presented on radio station WHA, Madison, 7 November 1962. Copy from Professor Aaron Ihde, Department of the History of Science.

4. William Darby, "A Scientist Looks at *Silent Spring*," *Chemical and Engineering News* 40 (1 October 1962), 60–63.

of the material that disrupted reproduction in birds, these uses added to the environmental load, which did cause harm. Since we could use other chemicals, he rejected Sweeney's conclusions and decided that we should stop using DDT, for our safety and the preservation of nature.

DDT use declined around the world in the 1970s, but only in part because of the American ban. Rising insect resistance and the availability of newer chemicals played important roles as well. It seemed by the 1990s that DDT was finally on the way out, but rising malaria rates in tropical countries stalled action on a worldwide ban, and that revived the continuing, low-level American debate over the environmental movement, with Carson and DDT becoming the occasion for arguments, like those over *Silent Spring*, over humans' relation to nature. The compound lingers, and the issue remains a touchstone for attitudes about nature and humans' relation to the world.

Books of readings include strong positions on both sides to generate good (or at least impassioned) classroom discussion, but they make the picture more clear-cut and polarized than it actually is. Despite the rhetoric, very few environmentalists worshipped trees, and none wanted to sacrifice people to save endangered toads. On the other side, that box of "Spotted Owl Helper" by Buzz Saw Productions—its logo a spinning saw blade with feathers and two owl eyes flying out—was indeed a joke and not a real food product. Chemical company executives belonged to the Sierra Club, environmentalists used modern medicine, and policy choices involved not conquering nature or living in harmony with it (both conditions difficult to define and impossible to reach) but balancing, with imperfect knowledge, different good things. DDT's story follows a common line in industrial civilization: we use new things because of their benefits and later find out their drawbacks. We live with nature and also from it, possibly not destroying but always changing it by our presence and actions.

QUESTIONS FOR THE DOCUMENTS

These documents raise many questions but only a few issues, and one good way to start exploring them is to ask what each author meant by key terms like "science," "nature," and "progress," or less general but still important ones like "safety," "health," and "toxicity." Take the word "nature." In English it meant, among other things, (a) everything in the universe, (b) everything outside human beings, (c) areas beyond settlement, (d) wholly "wild" areas, or (e) the essential character of something (as in "human nature"). Even when used to describe just plants and animals, it suggested everything from a world of wonders we needed for our psychic health to what we had to conquer to build civilization. That may seem far away from discussions of pesticides, but Stephen Forbes's discussion of the ecological foundations of economic entomology (see p. 15) and L. O. Howard's call for a "war against insects" (p. 20) depended on different views of humans' relation to nature. So did the blasts against *Silent Spring* in Part Four and the appreciation of it in Part Five. On a more concrete level, definitions of "safety" depended on the criteria used. Early studies of pesticide residues looked at the question with medical standards of illness in mind, while later ones drew on public health and were more concerned with chronic conditions than with what forced you to lie down for a few days. Still later ones extended "safety" to wildlife and ecosystems. Every document has definitions and views lurking in the background, and disagreements that seem related to facts often turn out to be tied to deep beliefs.

It is always tempting but usually misleading to divide the authors into the good guys and the bad guys. The first produces puzzlement (How could they have done that?), the second outrage (How could they have done that!), and neither helps us understand what went on. Better to ask what people knew or should have known at the time and what standards of harm they had in mind. Neal and his colleagues, testing DDT during World War II (see p. 26), never thought about long-term damage to the environment, and not just because their assignment was to measure medical effects on soldiers exposed to a few sprays. No one had ever used a synthetic organic chemical in such quantities or over so much land. But what should economic entomologists, ecologists, ornithologists, gardeners, or suburbanites have known in 1960, or 1970, or now? Ask as well what beliefs people shared, and on what level. Everyone wanted progress, but one kind led to chemical pesticides, another to biological control. Everyone used "science" to justify their stands, but they had different sciences in mind. Ask which one they liked and why.

PART 1

BACKGROUND

What people thought about nature and what they had for insecticides shaped their reception of DDT. The first two selections present two views of humans and nature, both well established in Western societies, through speeches by two eminent economic entomologists, made during the years when lead and calcium arsenate were becoming the pesticides of choice and the only alternatives—expensive and very temporary—were plant materials like pyrethrum, an extract from a flower grown mainly in Africa. One argued for control based on fundamental studies of insects and their ecology, including their interactions with humans, while the other, seeing an immediate crisis, argued that we needed chemicals now to win a "war against insects." Their stands came, at least in part, from their backgrounds. Stephen Forbes began as a medical student and, following service in the Illinois cavalry in the Civil War, became a naturalist and the head of the Illinois Natural History Survey (still in existence and now doing ecological research). Advising Illinois farmers at a time when effective chemical pesticides were not widely available, he became an expert on nonchemical insect controls, a stand perhaps made more emphatic by his own pioneering work in ecology. Leland O. Howard, a generation younger, also began in medicine but went into federal research at a time when materials and methods made chemicals much more effective.

"The War against Insects" came from the heart, for he titled his autobiography *Fighting the Insects.*

The next piece comes from the last major study of lead arsenate, a survey by the U.S. Public Health Service of some 1,200 people living in an apple-growing area in Washington State, done just a few years before DDT replaced it. This work used measures of safety that had guided early studies like the 1928 "Analyses of Sprayed Apples for Lead and Arsenic," measures based on sickness. That study reported the amounts of residues on apples "that had received the standard spray schedule comprising five applications of lead arsenate (four lbs. to 150 gals.), during the season of 1927." They found some lead and arsenic, but on the average levels were "considerably below the limit adopted by the [British] Royal Commission on Arsenical Poisoning in 1903 (1.429 mg. per kg.) [milligrams per kilogram] [and] no sample . . . exceeded that limit."[1] Early and late, this work assumed that materials stayed where they were applied, that the health of people occupationally exposed gave adequate warning of problems in the wider population, and that chronic, low-level exposure produced the same symptoms and problems as large doses.[2]

The next report comes from the wartime tests on DDT. Done by some of the same scientists involved in the lead arsenate work and designed to measure the effects of short-term exposures, it looked at acute conditions and symptoms under what were expected to be wartime conditions, and the research assumed exposure would end with the war. But as Paul Dunbar, an FDA scientist, admitted to a congressional committee investigating food and drug safety around 1950, wartime tests had been incomplete, and the decision to use DDT was "a calculated military risk" (a selection from one of Dunbar's speeches is on p. 51).[3] Public health workers could accurately estimate how many would die or be made ill if the sprays were not

1. Albert Hartzell and Frank Wilcoxon, "Analysis of Sprayed Apples for Lead and Arsenic," *Journal of Economic Entomology* 21 (February 1928), 125–30.

2. The medical literature shared these views. See "Dangers of Lead Arsenate as Spray in Orchards," *Journal of the American Medical Association* 105 (17 August 1935), 531; "Spray Residue on Foods" (editorial), *Journal of the American Medical Association* 108 (3 April 1937), 1178; "Hazards of Contaminated Fruits and Vegetables" (editorial), *Journal of the American Medical Association* 109 (10 July 1937), 135.

3. Testimony of Paul Dunbar in U.S. Congress, *Chemicals in Food Products*, Hearings before the House Select Committee to Investigate the Use of Chemicals in Food Products, 81st Cong., 2nd sess. (Washington: Government Printing Office, 1951), 236.

used, and against that toll the risks of unknown long-term damage seemed acceptable, particularly since most people exposed would be people in good health. What they could not calculate and did not know was that DDT behaved in the environment and people's bodies in a very different way than the arsenates and that animals, fish, and other creatures, as well as humans, would suffer from its use.

1

The Ecological Foundations of Applied Entomology

STEPHEN A. FORBES

Applied entomology is peculiarly an American subject, and here if anywhere in the world it should have accomplished its ends or should at least be in sight of its goal; and yet we have to acknowledge that, after generations of work upon them, many of the great long-standing problems of our American entomology are still unsolved, and that the people of our country are still suffering enormous losses of various description because of this fact. It is not because we do not know what we commonly call the *entomology* of the chinch-bug and the Hessian fly and the white-grubs and the cotton-moth that we are so nearly at our wit's end in our efforts to devise means for their control; it is because the knowledge of their entomology merely is not sufficient for the purpose. We are obliged to apply for assistance to the physiologist, and the chemist, and the physicist, and the meteorologist, and the geographer, and the agriculturist, and the animal husbandman, and the bacteriologist, and the physician, and the sanitarian, or, in a word, to the ecologist, who, from the nature of his studies, must, if he is thoroughly to cover his field, be something of each and all of these, and still something more.

The last and most essential phase in the expansion and development of our subject is the actual, practical, thoroughgoing *application* of the prod-

Annals of the Entomological Society of America 8 (1915), 1–19.

ucts of all our work. . . . Entomology which is not applied is not of its purposed end. . . .

This is an especially important point to us just now, for before we can discuss intelligently the foundations of applied entomology we must know how far the structure is to extend whose foundations we are about to plan. It is my insistent argument that it must, in the very nature of the case, cover the whole field of publication, education, community organization, cooperative effort, and legal compulsion necessary to give the fullest effect to the practical outcome of our entomological work; that our responsibilities, as official entomologists at any rate, do not end until we have done our best to see that all this is done or at least provided for. Just what this signifies with respect to the ecological foundations of applied entomology we shall be in better position to see when we have come to conclusions as to the meaning of ecology itself, and as to the general relations of that subject to entomology as actually applied. . . .

Let us agree, then, that, for the purposes of this discussion at any rate, the subject matter of ecology may be defined as the relation of organisms to their environment, and that this means the *whole* environment, organic and inorganic, and any and *all* organisms, man included—man, indeed, as by far the most important living factor, from whatever point of view. And let us also understand that the relations meant are, first, relations of *interaction*—*dynamic* relations, of efficient cause, and effect produced upon the organism by its environment and upon the environment by the organism; second, *space* relations, of distribution, position, juxtaposition, and association—static relations, we may call these, since they show the status of an individual or a group at a given time with reference to the various objects of its environment; and third, *successional* relations, time relations, sometimes called genetic because, in showing the static relations of a group in successive periods, they trace the genesis of the present status. . . .

Furthermore, there can be no doubt that it is primarily the dynamic factor only in ecology which interests the economic entomologist. It is only what insects do which gives them any importance, and it is only what can be done to them or about them in turn which gives applicable value to our knowledge of them and of their economy. We wish to know where they are or may be, how they are associated, and from whence they have come and by what they are likely to be succeeded, simply because their activities make them important to us. If they were inert we should not care. . . .

And now what shall we say of that view of ecology by which man, with his unrivalled powers of action and influence—the center and source of the most amazing interactions ever known between an animal species and its environment—is left practically outside the natural system, or is looked upon at best as a merely monstrous overgrowth of it—a *pathological* influence, a destructive enemy of nature, all whose works are *artificial* as compared with the *natural* effects and products of the vital activities of ants and caterpillars and crawfishes? There are ecologists to whom primitive nature is the earthly paradise, and civilized man is a kind of fiend, a Satan bent upon its destruction—a triumphant Satan who seems bound to reduce the whole earth, except, perhaps, the national parks, zoological gardens, bird preserves, and the like, to conditions as unnatural, as abnormal, as those of a prison or a hospital. Their ecology is a system not of this present time but of the world before Adam, before the fall of man had introduced into the world the germs of that fatal and frightfully contagious disease known as civilization. . . .

The ecological system of the existing twentieth century has man as its *dominant* species—dominant not in the sense of the plant ecologist, as simply the most abundant—for which idea *prevalent* would, I think, be a better term—but dominant in the sense of dynamic ecology, as the most influential, the controlling or *dominant* member of his associate group.

In applied entomology this is all of course very obvious, and needs no elaboration; for the economic entomologist is an ecologist pure and simple, whether he calls himself so or not—a student primarily of the interactions of insects and men, of that part of the actions and ecology of insects by which the welfare of man is affected, of that part of the ecology of insects which overlaps upon the ecology of man and that part of the ecology of man which overlaps, or can be profitably made to overlap, upon the ecology of insects. And it is the human interest which predominates and controls; the motive to applied entomology is primarily humanitarian. If there were no human interest to which entomology is applicable, there would be no applied entomology.

Now, since the field of applied entomology is precisely and solely that part of ecology in general over which the ecology of man and that of insects is coincident. . . , it must be evident, *a priori*, that a knowledge of the broad field of ecology as a whole, and of its general aims, principles, processes, and products, is fundamental to the special studies of the economic entomologist. It is only in some such sense as this that we can properly speak of the "ecological foundations of applied entomology" at all. The very substance of applied entomology being ecological through and through, it can have a foundation in ecology only as a part of a whole. . . . It is my special task, therefore, to point out and illustrate some of the ways in which general ecology may be made helpful to applied entomological ecology, and, *vice versa*, ways in which applied entomology may be made useful—is already useful, indeed—to the student of general ecology. . . .

How many of our measures of protection and defense against insect depredations depend upon any precise knowledge of general fact or scientific principle, or are traceable to anything better than a purely empirical warrant? If we attempt to analyze what we know and what we still need to know concerning any one of the great insect pests before we shall be in a position to do all that can be done and ought to be done to restrain its ravages and injuries either by measures of avoidance, prevention, mitigation, or arrest, we may perhaps get a clearer, concrete idea of what is involved in economic entomology, and what are the foundations of fact and principle upon which it rests. . . .

We know the ordinary life history of the chinch-bug fairly well, although our knowledge is still lacking in the details of variation of life history in different regions, seasons, and climates; while of its so-called physiological life history we know almost nothing exact. . . .

We know that certain insecticide substances in solution or emulsion are effective against it in a way to make them practically available, but we do not know how or by what properties they produce their fatal effects and we are consequently without definite guidance in our search for other such insecticides.

We know that any and all measures against this insect are of comparatively little avail if undertaken sporadically, by an individual only here and there; that for their fair and full effect they must be made the fixed policy and practice of whole communities, actuated by the community motive as well as the personal one. We know indeed that a large part of our applied entomology fails of its application because communities are not brought to the point of cooperative action in the general interest; but we do not know—we have scarcely discussed among ourselves—the best means of appeal and the best methods of organization and management to effect these results, without which much of our economic entomology must fall practically short of the economic end. . . .

If you ask me now whether we should be any nearer the practical control of our most dangerous and destructive insect pests if we had the details of their ecology well worked out, I shall have to answer that I do not know, any more than the entomologists who studied the habits and general ecology of mosquitoes foresaw the use of their observations as an indispensable link in the study and control of malarial disease—any more than Laveran knew when he found a blood parasite associated with malarial disease in man that the remaining links in the chain would presently be traced. . . .

To me it seems so evident that such a knowledge would be of the greatest value to the investigating economic entomologist, that I am quite prepared to [say] that the greatest need of applied entomology at the present time is just this kind of scientific ecology, and that it is among our first and most important duties to acquaint ourselves with this field and to encourage, provide for if possible, and assist as we can, serious, exact, and thoroughgoing work in scientific entomological ecology. . . .

If applied entomology is essentially a mixture of human and insect ecology, then it seems clear that courses in general ecology should form a part of the education of the economic entomologist. Indeed, I have much tangible evidence of the value of this combination in the results shown in my own university department of entomology, whose more capable students all tell me of the unique advantage which they find in ecological courses because of the broader outlook and the new point of view which these give them, and especially because of the greater theoretical interest of their technical studies when related to the foundation principles of ecology. . . .

I believe that students of ecology itself would be equally, although somewhat differently, profited if they were to take one or more economic courses in entomology; that they too would find a new outlook thrown open to them and a new and larger meaning given to their work. I hope that the time may soon come when ecology shall be taught in at least every state university and every agricultural college, and when something of applied biology shall be included among the regular courses of every university student specializing in ecology. Then for the first time we may be in a position to estimate fairly the value of the contributions which entomological ecology, fully and thoroughly applied, may be competent to make to the progress of biology and to the welfare of civilized man.

2

The War against Insects

LELAND O. HOWARD

The war against insects has in fact become a worldwide movement which rapidly is making an impression in many ways. In the United States, for example, investigations in this field are for the time being receiving the largest government support. Every state has its corps of expert workers and investigators. The Federal government employs a force of four hundred trained men and equips and supports more than eighty field laboratories scattered over the whole country at especially advantageous centers for especial investigations. And there are teachers in the colleges and universities, especially the colleges of agriculture, who are training clever men and clever women in insect biology and morphology and in applied entomology both agricultural and medical.

All this means that we are beginning to realize that insects are our most important rivals in nature and that we are beginning to develop our defense.

While it is true that we are *beginning* this development, it is equally true that we are only at the start. Looking at it in a broad way, we must go deeply into insect physiology and minute anatomy; we must study and secure a most perfect knowledge of all of the infinite varieties of individual development from the germ cell to the adult form; we must study all of the aspects

Chemical Age 30 (January 1922), 5–6.

of insect behavior and their responses to all sorts of stimuli—their tropisms of all kinds; we must study the tremendous complex of natural control, involving as it does a consideration of meteorology, climatology, botany, plant physiology, and all the operations of animal and vegetal parasitism as they affect the insecta. We must go down to great big fundamentals.

THE WORK OF THE CHEMIST IN THE WARFARE AGAINST CROP PESTS

All this will involve the labors of an army of patient investigators and will occupy very many years—possibly all time to come. But the problem in many of its manifestations is a pressing and immediate one.

That is why we are using a chemical means of warfare, by spraying our crops with chemical compounds and fumigating our citrus orchards and mills and warehouses with other chemical compounds, and are developing mechanical means both for utilizing these chemical means and for independent action. There is much room for investigation here. We have only a few simple and effective insecticides. Among the inorganic compounds, we have the arsenates, the lime and sulphur sprays, and recently the fluorides have been coming in. Of the organic substances, we use such plant material as the poisons of hellebore and larkspur, pyrethrum and nicotine; and the cyanides and the petroleum emulsions are also very extensively used. No really synthetic organic substances have come into use. Here is a great field of future work.

Some of the after happenings of the war have been the use of the army flame-throwers against the swarms of locusts in the south of France, the experimental use against insects of certain of the war gases, and the use of the aeroplane in reconnaissance in the course of the pink bollworm work along the Rio Grande, in the location of beetle-damaged timber in the forests of the Northwest, and even in the insecticidal dusting of dense tree growth in Ohio. The chemists and the entomologists, working co-operatively, have many valuable discoveries yet to make, and they will surely come.

All this sort of work goes for immediate relief. Our studies of natural control follow next. It is fortunately true that there are thousands upon thousands of species of insects which live at the expense of those that are inimical to man and which destroy them in vast numbers; in fact, as a distinguished physicist in discussing this topic with me recently said, "If they would quit fighting among themselves, they would overwhelm the whole vertebrate series." This is in fact one of the most important elements in natural control and is being studied in its many phases by a small but earnest group of workers.

BIOLOGICAL CONTROL

So far, while we have done some striking things in our efforts at biological control, by importing from one country into another the natural enemies of an injurious species which had itself been accidentally introduced, and while we have in some cases secured relief by variations in farm practice or in farm management based upon an intimate knowledge of the biology of certain crop pests, we are only touching the border of the possibilities of natural control.

For an understanding of these possibilities, we must await the prosecution of long studies, just as we must await years of progress of those other studies outlined in a previous paragraph. And all of these studies must be carried on by skilled biologists—thousands of them. At present most of the best men are working away in their laboratories practically heedless of the great and inviting lines of study at which I have hinted and heedless of the tremendous necessity of the most intense work by the very best minds on the problem of overcoming and controlling our strongest rivals on the planet.

THE WARFARE OF NATURE

Let us summarize. Few people realize the critical situation which exists at the present time. Men and nations have always struggled among themselves. War has seemed to be a necessity growing out of the ambition of the human race. It is too much, perhaps, to hope that the lesson which the world has recently learned in the years 1914 to 1918 will be strong enough to prevent the recurrence of international war; but at all events, there is a war, not among human beings, but between all humanity and certain forces that are arrayed against it.

Man is the dominant type on this terrestrial body; he has overcome most opposing animate forces; he has subdued or turned to his own use nearly all kinds of living creatures. There still remain, however, the bacteria and protozoa that carry disease and the enormous forces of injurious insects which attack him from every point and which constitute today his greatest rivals in the control of nature. They threaten his life daily; they shorten his food supplies, both in his crops while they are growing and in such supplies after they are harvested and stored, in his meat animals, in his comfort, in his clothing, in his habitations, and in countless other ways. In many ways they are better fitted for existence on the earth than he is. They constitute a much older geological type, and it is a type which had persisted for countless years before he made his appearance, and this persistence has been due to characteristics which he does not possess and can not acquire—rapidity of multiplication, power of concealment, a defensive armor, and many other factors contribute to this persistence.

AN IMPORTANT FIELD OF RESEARCH

With all this in view, it will be necessary for the human species to bring this great group of insects under control, and to do this will demand the services of skilled biologists—thousands of them. We have ignored these creatures to a certain extent on account of their small size, but their small size is one of the great elements of danger, and is one of the great elements of success in existence and multiplication.

Let all the departments of biology in all of our universities and colleges consider this plain statement of the situation, and let them begin a concerted movement to train the men who are needed in this defensive and offensive campaign.

In closing, I cannot refrain from quoting a remarkable paragraph from Maeterlinck:

> The insect does not belong to our world. The other animals, even the plants, in spite of their mute existence and the great secrets which they nourish, do not seem wholly strangers to us. In spite of all, we feel with them a certain sense of terrestrial fraternity. They surprise us, even make us marvel, but they fail to overthrow our basic concepts. The insect, on the other hand, brings with him something that does not seem to belong to the customs, the morale, the psychology of our globe. One would say that it comes from another planet, more monstrous, more energetic, more insensate, more atrocious, more infernal than ours. . . . It seizes upon life with an authority and a fecundity which nothing equals here below; we cannot grasp the idea that it is a thought of that Nature of which we flatter ourselves that we are the favorite children. . . . There is, without doubt, with this amazement and this incomprehension, and I know not what of instinctive and profound inquietude inspired by these creatures, so incomparably better armed, better equipped than ourselves, these compressions of energy and activity which are our most mysterious enemies, our rivals in these latter hours, and perhaps our successors.

3

A Study of the Effects of Lead Arsenate Exposure on Orchardists and Consumers of Sprayed Fruit

PAUL NEAL ET AL.

This investigation comprises a study extending over three years of the possible injury to health of people exposed to lead arsenate whether by ingestion on fruit (consumers), by inhalation of spray mist or dust (orchardists), or by other forms of exposure. It has included both an intensive field study of large groups of men, women, and children, as well as toxicologic laboratory investigations of the effect of lead arsenate upon man and animals, and of similarly-related problems. The possibility that lead arsenate could have been a factor, directly or indirectly, in other diseases was studied.

This bulletin reports the results of an epidemiologic study based on field operations extending over a fourteen month period of 1,231 men, women, and children who live in an apple-growing region where large quantities of lead arsenate have been used for many years as insecticide sprays. Toxicologic studies of the effect of lead arsenate on man and laboratory animals are being published separately. . . .

Only six men and one woman had a *combination* of clinical and laboratory findings directly referable to the absorption of lead arsenate. Some

U.S. Public Health Service Bulletin 267 (Washington: Government Printing Office, 1941). Paul Neal, Waldeman C. Dressen, Thomas I. Edwards, Warren H. Reinhart, Stewart H. Webster, Harold T. Castenberg, and Lawrence T. Fairhill.

physicians may interpret these cases as minimal lead arsenate intoxication. However, as regards lead, these cases do not come up to the criteria of the Committee on Lead Poisoning of the American Public Health Association for lead intoxication, incipient plumbism or lead poisoning. These subjects were all orchardists and ranged in age from twenty-three to sixty-eight years.

No one else had a combination of clinical and laboratory effects directly attributable to lead arsenate absorption. These include: ninety-five men and 146 women classified as consumers who have no occupational exposure to lead arsenate or to any other lead or arsenic compound; 158 men and 171 women, former orchardists or seasonal workers in apple-packing sheds, who are intermediate in exposure between the two foregoing groups; and ninety-nine boys and girls under fifteen years of age. . . .

Special attention was given to medical examination of children because, in this district, where orchards surround the communities or the houses in which they live, there are unusual opportunities for children to be exposed to lead arsenate insecticide sprays and spray residues on branches, leaves, and grass, in addition to the lead arsenate insecticide residues they ingest on apples. There was only one respect in which these children may have differed from children in other districts; their urinary lead and urinary arsenic values were higher than the corresponding values for a group of eighteen children living near Washington, D.C., known to have no unusual exposure to lead or arsenic.

There was no indication of adverse effects of lead arsenate exposure on the health of these children. . . .

4

Toxicity and Potential Dangers of Aerosols, Mists, and Dusting Powders Containing DDT

PAUL NEAL ET AL.

The following study of the toxicity of DDT when used as an aerosol, dusting powder, and as a spray in deobase was undertaken upon the request of the Surgeon General's Office of the United States Army. . . .

The acute toxicity of various DDT aerosols was studied in mice, rats, guinea pigs, and dogs. The animals were kept in wire cages of appropriate size and were exposed in a glass chamber of 409.7 liters capacity. The aerosol was introduced into this chamber through a hole in the bottom by means of an adapter screwed onto the aerosol cylinder, and the animals were left in the static atmosphere of the chamber for a total of forty-five minutes, a period sufficiently short to prevent the accumulation of excessive amounts of carbon dioxide or an appreciable reduction of the oxygen content of the air. Approximately one pound of the DDT aerosol was discharged into the chamber at the beginning of the experiment. This took about twenty minutes. . . .

Human subjects tolerate exposure for one hour daily, on six consecutive

Supplement 177 to *U.S. Public Health Service Reports* (Washington: Government Printing Office, 1944). P. A. Neal, W. F. von Oettngen, W. W. Smith, R. B. Malmo, R. C. Dunn, H. E. Moran, T. R. Sweeney, D. W. Armstrong, and W. C. White.

days, to a dispersion of DDT in air prepared by introducing 10.4 gm. of an aerosol containing 5 percent DDT, 10 percent cyclohexanone, and 85 percent Freon, every fifteen minutes, into a sealed chamber of 14,750-liters capacity without any untoward effects except a moderate irritation of the eyes and the upper respiratory tract. Exposure for one hour daily, on five consecutive days, to a dispersion of the same aerosol prepared by introducing 10.4 gm. every five minutes into the same volume of air also failed to produce any systemic toxic effects in human beings.

The experiments described in this report allow the following conclusions:

In spite of the inherent toxicity of DDT, its use in a 1 to 5 percent solution in 10 percent cyclohexanone with 89 or 85 percent of Freon as aerosol should offer no serious health hazards when used under conditions such as those required for its use as an insecticide. It should be pointed out that the solution of DDT in fatty oils definitely increases its toxicity, and that the results obtained by using a solution of DDT in cyclohexanone are not necessarily comparable to the effects produced by a solution in oil.

The use of DDT in concentrations up to 10 percent in inert powders, for dusting clothes, as in the extermination of lice, appears to offer no serious hazards because of the relative insolubility of DDT and the large particle size of the dust. Therefore, it does not reach the alveolar spaces. A large proportion of the dust is retained in the uppermost sections of the respiratory tract. The remainder is swallowed. On account of its relative insolubility it is thought that only a small fraction is absorbed.

Because the use of a 1 percent DDT-deobase mixture was found to be nontoxic to rabbits with heavy exposure for forty-eight minutes daily, over a period of four weeks, it is believed that its use as a fly spray, which involves only temporary and comparatively moderate exposure to much lower concentrations, should be safe. However, due to the fat-solvent properties of most petroleum distillates, irritation of the skin may occur following heavy exposure.

Although this study deals only with the appraisal of the potential dangers of DDT when inhaled as aerosol, dust, and mist, it should be pointed out that ingestion of massive doses of DDT will cause a toxic reaction. It should, therefore, only be used under conditions which exclude the heavy contamination of food.

Since these experiments were concluded a thorough clinical and laboratory study has been made of three men who have each had several months continuous occupational exposure to DDT used in various forms as an insecticidal agent.

An evaluation of the results of these examinations fails to indicate any definite evidence of toxic effects from the exposure the three subjects have had to DDT.

PART 2

DDT'S BRIGHT PROMISE AND NEGLECTED PROBLEMS (1942–1958)

The public first heard about DDT as the wartime miracle chemical that saved our troops and European refugees from disease and encountered it after the war as a household wonder and agricultural marvel. For a decade only two groups paid much attention to it: the economic entomologists who recommended sprays and the public health officials who regulated residue levels on food. Wildlife biologists were interested but had no influence on policy and did not argue for it. The first piece comes from January 1945, when the war was still raging in Europe and the Pacific. In "How Magic is DDT?" Brigadier General Simmons, an Army officer who spent thirty years working on public health problems in bacteriology and tropical medicine, described the chemical's great promise in controlling tropical diseases but also its limitations. The next piece comes from months just after the war, as manufacturers and formulators began to adapt the new material to a civilian market. "Aerosol Insecticides" shows the adaptation of the wartime miracle to civilian use, as well as the application of precautions taken from military guidelines. The advertisement "'Do's and Don'ts' in Household Insecticide Application" was part of the industry's campaign to introduce the public to the new compound and to highlight the industry's reliance on military technology and experience as well as the concern about safety and liability. Look at these from the point of view of 1945. The industry followed standard practices in spraying and dusting millions, and while its assumptions about the safety and promise of chemicals and a confidence in technology now seem as odd as the women's clothes and hairstyles, they followed what was known then.

The next three selections reveal several sides to the questions of safety and effectiveness. Clay Lyle's presidential address to the American Association of Economic Entomologists, "Achievements and Possibilities in Pest Eradication," calling for a new "war against insects," should be read with the speeches of Forbes and Howard from Part One in mind, because—besides their views about humans and nature—all deal with the cooperation and even legal coercion needed to make insect control programs effective. Paul Dunbar's talk to the National Agricultural Chemicals Association (the manufacturers' organization), "The Food and Drug Administration Looks at Insecticides," came from a professional lifetime as a government scientist dealing with chemicals and human health in the Bureau of Chemistry and its successor agency, the Food and Drug Administration. His argument for caution rested on the limitations of wartime tests and discoveries made since the war. Clarence Cottam and Elmer Higgins came from a different direction. They were Fish and Wildlife Service biologists, and their report, "DDT and Its Effect on Fish and Wildlife," summarized the agency's studies, begun in 1944, and other work on wild and agricultural land. They published it, reasonably enough, in the *Journal of Economic Entomology*, read by the people responsible for recommending sprays. It might seem an early "environmental" piece, but it looked not at sprays but the misuse of sprays, and only ecologists knew about the "environment" in 1946. Here again, ask what they knew but also what they could not have known.

5

How Magic Is DDT?

BRIGADIER GENERAL JAMES STEVENS SIMMONS

The initials DDT are used as a convenient nickname for a jawbreaking chemical term, dichloro-diphenyl-trichlorethane. In everyday language, this high-sounding chemical compound is a stable, almost colorless and practically odorless crystalline solid.

It is not soluble in water, but can be dissolved in many organic solvents, including kerosene and various other oils. It is one of the most powerful insect poisons known, one which affects the nervous system of the insect and produces jittery, spasmodic movements, followed by paralysis and, later, by death.

It is effective when used in infinitely small amounts and can be used highly diluted either in a powder or in oily solutions. It can kill many of our innumerable insect enemies, not only the annoying household pests and many of the plant parasites that ravage our crops and food supplies but also lice, mosquitoes and other dangerous, blood-sucking insects which are responsible for the spread of typhus fever, malaria and other serious diseases.

DDT is of great importance to all of us, both in helping to win the war and improving the country's health after the war is over. DDT itself is fairly old. It has been known for seventy years, and its ability to kill certain of the insect pests of plants was observed about four years ago in Switzerland. However, its value as a military weapon for the control of insect-borne dis-

Saturday Evening Post 217 (6 January 1945), 18 ff.

eases has been discovered and developed by scientists in this country only during the past two years. At first the results of the experimental work were blanketed by military secrecy. Recently, however, this secrecy has been lifted; the story of the use of DDT to control typhus in Naples has been announced and statements have been released concerning its effective use against malaria in many theaters.[1]

Such reports have fired the popular imagination, and the symbol DDT is acquiring a mysterious, romantic aura. It is coming so rapidly into common use that it bids fair to join the ranks of such well-known war-born Army terms as "jeep," "radar," and "bazooka."

As the experimental work with DDT continues at top speed both in the laboratories and in the field, at home and abroad, the reports of progress which reach the Surgeon General's Office almost daily compete in interest with the war bulletins from the fighting fronts. Meanwhile these reports of the amazing uses of DDT are passed over for yarns telling of its destructiveness which sound like newly created versions of the Arabian Nights. These incredible rumors picture DDT as a substance which may bring complete ruin to both the animal and the vegetable kingdoms. For example, a serious scientific report that DDT has killed millions of malaria mosquito larvae in Gatun Lake may be overshadowed by a fantastic story claiming the particles of the chemical, transported by the trade winds, have annihilated all the blue butterflies in the Isthmus of Darien. . . .

Many uninformed persons have been puzzled as to why such an old chemical as DDT was not adopted earlier by the Army. The answer is that while this chemical, like thousands of others, has been known for many years, the methods evolved for its use by the armed services are entirely new. They have been developed with phenomenal speed as a part of a streamlined program of wartime medical research. As soon as DDT's effectiveness and safety were established, no time was lost in putting it into use. . . .

In the fall of 1942, a small amount of Gesarol[2] was sent by the parent company to its New York office, along with a report that it had been useful in saving the Swiss potato crop from invasion by the Colorado potato beetle, and also in the destruction of various local plant pests. It was claimed

1. Correspondent Allen Raymond told the Naples story in "Now We Can Lick Typhus" in the *Post* for April 22, 1944. Censorship restrictions in effect at the time prevented him from mentioning DDT by name. —*Post* ed.

2. The Geigy company's name for its formulation.

that when painted on the walls of a barn, a single application would kill any flies that chanced to light there for a period of a month or more thereafter.

This sample reached the United States at an opportune time, for the Army was then looking for a new insecticide. On arrival, however, its chemical identity was still hidden under the Swiss trade name Gesarol and the Army had no information about its value against the blood-sucking insect parasites of man. Nevertheless, because of the fly story, it was welcomed as another preparation of sufficient promise to be tested for its potential value to the armed forces. Thus, recognition of its full possibilities might have been delayed indefinitely, had it not been for the problems posed by a global war.

The Army needed another insecticidal substance, because the war had cut off the normal supplies of rotenone from the Dutch East Indies, and an acute shortage of pyrethrum threatened to cripple its program for the prevention of insect-borne diseases. . . . Many American scientific agencies were mobilized to assist in this search. . . .

As a result of this comprehensive research program, a series of excellent new insecticidal agents had been produced for the armed forces. These agents included methyl-bromide gas, which is used throughout the Army as a convenient fumigant for the rapid delousing of clothing and equipment; a louse powder known as MYL, used until recently to delouse the individual soldier; three valuable insect repellents used to protect soldiers against the bites of mosquitoes, sandflies and chiggers; and the popular Freon-pyrethrum mosquito bomb, which is used by our troops in all tropical theaters to kill adult mosquitoes.

While the War Department was making strenuous efforts to speed the production and distribution of these new insecticides to the troops, the supply of the Army louse powder and the mosquito bomb, both of which required pyrethrum, was jeopardized by a combination of temperamental weather conditions, food crop failure and labor difficulties in far-away Kenya colony on the east coast of Africa. These unpredictable circumstances had reduced the crop of pyrethrum, which is a daisylike flower of the chrysanthemum family. As Kenya was the chief source of our national supply, this accident threatened to cut off the annual flow of millions of pounds of this essential insecticidal material to the United States.

An attempt was made to meet the pyrethrum shortage by conserving the supplies remaining in the country for the armed forces alone, and an unsuccessful effort was made to obtain pyrethrum seeds from Kenya for planting in the Western Hemisphere. However, the supply continued to dwindle and there was an increased demand for it, not only from agriculture but from the troops of the United Nations. Soon it was necessary to ration the Army louse powder and to limit the use of the Army mosquito bomb to our overseas tropical theaters and to the disinsectization of military aircraft. The situation became so serious that a substitute for pyrethrum had to be found.

The Army sent out an SOS to its cooperating research agencies, and an immediate search was begun. Many previously disregarded chemicals were re-examined, including some that formerly had been considered unsuitable for military use. During this critical period, DDT arrived in America and began its remarkable wartime career.

The Gesarol sample was examined by experts of the Department of Agriculture, who found that it contained DDT, which they resynthesized and used in insecticidal studies in the laboratories of the Bureau of Entomology and Plant Quarantine at Orlando, Florida. The Swiss reports of its action on plant pests were found to be true, and our scientists immediately began to study the effect of DDT on various other insects. The results were astounding. Not only was the prolonged killing effect on flies confirmed, but it was discovered that even when used in unbelievably small amounts the material had a similar action against mosquitoes, lice, fleas, bedbugs and other insect parasites that feed on man.

These studies showed that DDT acts as a nervous-system poison and that it kills certain insects either when swallowed by them or when brought into contact with them. For example, when mosquitoes touch an oily solution of DDT, they show no signs of being poisoned for about twenty minutes. Then they become nervous and agitated, take off abruptly, fly about in erratic, drunken circles and finally, after a binge of five to twenty minutes, drop to the floor. Paralyzed and unable to fly again, they die several hours later. These reactions are so typical that laboratory workers often refer to them as the "Gesarol jitters" or the "DDT's." The knockdown effect of DDT is slower than that of pyrethrum, but it retains its lethal power for a much longer time.

DDT is also highly poisonous to lice, and when mixed with certain inert powders, it provides a louse powder infinitely better than the MYL powder previously used by the Army. The MYL powder, when shaken into the clothes of a soldier, keeps him free from lice for about a week, but a single application of DDT louse powder is effective for more than a month. During field tests made with this powder among louse-infested natives in various parts of the world, it has been so popular that the investigators were frequently embarrassed by the large numbers of volunteers who demanded attention.

The toxic action of DDT is so strong that some of the scientists who first used it ruined important experiments because they failed to clean their insect cages before using them again, and the small amounts of DDT remaining were sufficient to kill the new insects introduced.

The remarkable results obtained during the early part of 1943 stimulated the hope that DDT might be immediately adopted for use in the Army, thus affording a way out of the pyrethrum dilemma. But before this could be done, a disturbing question had to be answered: Could DDT be used on the millions of soldiers and sailors of our armed forces with safety? The preliminary safety tests, made with full-strength DDT, had been somewhat alarming. When eaten in relatively large amounts by guinea pigs, rabbits and other laboratory animals, it caused nervousness, convulsions or death, depending on the size of the dose.

Further investigations were therefore required to determine its safety for man when small amounts were applied over long periods of time and under the exact conditions of its proposed military uses. It was soon learned that, used as a louse powder, DDT can be safely applied to the human skin, but it was not until the end of the year that it was shown to be safe for use in certain other ways. We now know that, when properly diluted, DDT can be combined with oil and used in sprays without danger from inhalation. We also know that, when used correctly in small amounts as a mosquito larvicide, it can be added to water without killing fish or game. It must be remembered, however, that like other insecticides, DDT is a powerful poison and has to be used intelligently. It must not be swallowed, and oily solutions of it must not be applied to the skin.

Announcement of the results of the final safety tests late in 1943 precipitated an enormous demand for DDT to satisfy the needs of our Army and Navy and the military and civilian requirements of the United Nations. In an attempt meet these important and sometimes unreasonable demands, the Headquarters Army Service Forces, the Surgeon General, the Quartermaster General and the War Production Board inaugurated a tremendous production program. Within a short time, the supply has increased enormously, but so have the demands. Unfortunately, the production is not yet adequate to meet civilian requests.

Our primary interest in DDT has naturally been concerned with its value as a new weapon for use in the field of military preventive medicine and public health. The addition of this insecticide to our medical arsenal has forged another vital link in the Army's chain of defense against disease and has materially strengthened the Surgeon General's program for protecting the health of our far-flung forces. . . .

The number one hazard of this war is malaria, which has been a serious problem in every tropical theater. Another exotic, insect-borne disease which is not tropical, but was feared at the beginning of the war because of its notorious past, is typhus fever. Armed with DDT, the Army has conquered the fear of typhus. For the first time in history, this ruthless companion of disaster, famine and poverty has lost all right to its murderous title of champion of the ancient plagues of war.

Army preventive medicine is also smashing ahead on all fronts in the stubbornly resisted fight against malaria. Both typhus and malaria are charter members of the ancient order of wartime diseases, but they differ in their techniques and spheres of influence. Typhus prefers cold and temperate climates, and it burrows about in its filthy endemic lairs until disaster affords a chance to attack; then, through its loathsome intermediary, the louse, it preys on the miserable and the weak. Malaria, on the other hand, is a disease of hot, humid regions, which strikes out directly; its fleets of airborne mosquito carriers hit with the fury of robot bombs, attacking both the weak and the strong.

It wages a continuous offensive against man throughout its great tropical domain which encircles the earth on both sides of the Equator. During the warmer months of the year it expands the borders of its empire as far north as Southern Sweden and Lake Ladoga; and as far south as Johannesburg and Argentina and, in rare instances, to Queensland. A complete world census of its victims has never been made, but in 1932 more than 17,000,000 patients were treated in sixty-five countries. In India alone, only one tenth of the annual toll of 100,000,000 malaria cases received treatment, and several million people died.

Moreover, chronic malaria has weakened and enslaved a large part of the earth's population and has played an important and often decisive role in most of the great wars fought within its jealously guarded empire. It is claimed that malaria was largely responsible for the decline of the ancient Greek and Roman civilizations. Alexander the Great, dreaming of new worlds to conquer, was cut off in his mighty youth by what was probably the plasmodium of malignant malaria slipped into his skin through the bite of a vagrant anopheline mosquito. . . .

The gigantic antimosquito campaign which has been carried on by the Army on military reservations . . . has reduced the prevalence of malaria among soldiers in the United States to the lowest rate ever recorded.

But the battle overseas has been a tough one. In certain tropical areas, the malaria rates during some of the early campaigns were excessive because of the unusual exposure of combat troops, difficulties in supplying them with malaria-control materials, and, in some instances, failure of commanders to appreciate the importance of enforcing sanitary discipline. During the last year this situation has been strikingly improved, and considered as a whole, the prevalence of malaria in the Army is not alarming. In June 1944, the American minister to Australia, Mr. Nelson Trusler Johnson, following a visit to New Guinea, wrote the Surgeon General that "amazing things have been accomplished, so that malaria is no longer the threat to success of our mission in this area of war."

The insect repellent which is now supplied to troops in all theaters repels mosquitoes for about four hours and is a boon to the soldier exposed at night under combat conditions. Mosquito-proof clothing impregnated with various repellents is being tested for use in certain regions. The spray of the Freon-pyrethrum bomb is used to kill adult mosquitoes in planes, barracks, tents and foxholes.

These measures for individual control are now being supplemented by DDT. Solutions in oil may be used as sprays. Applied to the walls of buildings or tents, such solutions, for periods of several weeks, kill all adult mosquitoes that light on them.

DDT is also so effective as a mosquito larvicide that it is replacing the methods formerly used. When a small amount of oily solution is dropped on the water at the edge of a pond, it spreads rapidly over the entire surface and kills all the mosquito wrigglers present. One pound of DDT in a 5 percent solution of Diesel or fuel oil is sufficient for five acres of water.

Examples have been reported of ducks that, after swimming on water treated with DDT, moved on to untreated ponds and carried on their feathers enough of the chemical to kill the mosquito larvae in the second pond. The most exciting development with DDT has been its experimental distribution from airplanes in the form of smokes and sprays to destroy mosquitoes in large inaccessible areas. Preliminary tests with slow-flying planes gave excellent results; both adult and larval mosquitoes were killed in areas covered.

This led to field experiments with fast combat aircraft which are now in progress in every tropical theater. The reports from all these tests indicate that, as already anticipated, DDT is the greatest weapon now available for continuing the fight of the armed forces against malaria.

During the last few years, many new agricultural uses of DDT have been discovered and its great military value for the control of typhus and malaria has been demonstrated. However, we have only scratched the surface of its potentialities. We now know that DDT is effective against the Japanese beetle, various cabbage worms, the coddling moth and many other destructive plant pests, but poor results have been obtained in experiments with the Mexican bean beetle, the red spider and the cotton boll weevil. It can be used to kill such domestic pests as flies, ants and cockroaches, although the German cockroach is more resistant than the American.

Also it is useful against many biting insects, including lice, mosquitoes, flies and bedbugs, and, to some extent, against ticks and chiggers. To those of us who believe that, when a magic key to world health is discovered, this key can work only in a lock well lubricated with the magic oil of world health, these potentialities are interesting. Therefore, the increasing production of DDT continues at full speed and the program of experimentation is being intensified and expanded in order to obtain answers to the innumerable questions which arise concerning its future uses. . . .

It is fully realized that such a powerful insecticide may be a double-edged sword, and that its unintelligent use might eliminate certain valuable insects essential to agriculture and horticulture. Even more important, it might conceivably disturb vital balances in the animal and plant kingdoms and thus upset various fundamental biological cycles. In order to investigate all phases of these broader problems as well as to give additional help to the armed forces during the present emergency, an important new board on insect control has recently been established by the Office of Scientific Research and Development.[3]

The possibilities of DDT are sufficient to stir the most sluggish imagination, but even if all investigations should cease today, we already have a proud record of achievement. In my opinion it is the war's greatest contribution to the future health of the world.

3. The Office of Scientific Research and Development coordinated war-time research, including the Manhattan Project, which produced the atomic bomb.

6

Aerosol Insecticides

Paralleling the tremendous wave of interest which the trade has shown in DDT since government restrictions were removed a few weeks ago, has come a comparable though perhaps less widespread demand for more information on civilian market possibilities for aerosol insecticides, another important wartime insecticide development. A number of very substantial factors are in various stages of progress with a number of interesting consumer aerosol packages, and while there seems to be little prospect of wide developments in the field this season, since the northern fly and mosquito season is about over, indications are that by next spring there will be a number of tested products on the market. With DDT, there has been a rush to market, and many of those who have rushed fastest have been firms who seem to know little or nothing about the product—or even about insecticides of any type. With aerosols, the interested firms seem in general to be well established companies who want to complete their test work, and to iron out all of the kinks in their products, before going to market.

The first aerosol products to reach the test marketing stage resemble, as they might be expected to do, the aerosol bombs used by the army and navy. They make use of DDT and highly concentrated pyrethrum extract as the

Soap and Sanitary Chemicals 21 (October 1945), 124–26.

"Do's and Don'ts" in
HOUSEHOLD INSECTICIDE APPLICATION

A SERIES of photographs showing correct and wrong technique in spraying household insecticides has recently been prepared by Hercules Powder Co., Wilmington, manufacturers of "Thanite," synthetic insecticide raw material, and *Soap & Sanitary Chemicals* is privileged to release the accompanying selected shots from the Hercules picture file. Some of the shots illustrate common errors in insecticide application, while others present approved spraying technique.

Other picture stories based on the Hercules photos are to appear subsequently in national magazines as a step to help educate the housewife in using insect sprays correctly. Hercules Powder Co. has also announced that insecticide manufacturers using Hercules insecticide raw materials in their formulations will be permitted to borrow and use individual shots from the Hercules picture file in their advertising and promotion pieces.

A similar set of pictures has been prepared covering the application of cattle sprays, and is now reported ready for distribution.

134 SOAP *and* SANITARY CHEMICALS April, 1946

Advertisement in *Soap* magazine, 1946

toxicants, continue to employ "Freon" as the dispersing gas, but have generally eliminated oil of sesame from their formulas. The container or dispenser is a close replica of the "bombs" supplied to the Army and Navy. The new civilian aerosol bombs are priced anywhere from $2.00 to $4.00—the

Advertisement in *Soap* magazine, 1946

latter top level just having been fixed by the OPA [Office of Price Administration] as a price ceiling.

As to specific formulation— one new product contains about 3 percent of a 20 percent solution of pyrethrins, 2 percent DDT, sufficient solvent to dissolve the toxic ingredients, and the balance is "Freon." For a time at least, it is doubtful if the DDT content will be boosted any higher than 3 percent. The United States Public Health Service has sanctioned the use of 3 percent

DDT in connection with the Army and Navy formulas for aerosol insecticides, and those adhering to this maximum in their formulations would seem to be on fairly safe ground in defending against any possible damage suits, as they would be able to defend their position on the basis of government tests and usage of 3 percent DDT by the Armed Forces. The pyrethrum content will be determined by price factors, as this is the highest cost item entering into the formula. The maximum effect is obtained, it is reported, by using 4 percent of a 20 percent pyrethrin concentrate, but price factors will reduce this content down to 3 or even 2 percent in different products. . . .

Westinghouse Electric & Manufacturing Company, who made so many millions of aerosol bombs for use by the Armed Forces, is perhaps as well along with plans for developing the civilian market as anyone else in the field. It is reported that they will run a market survey in Jacksonville, Fla., next month and will sell 10,000 of their one-pound containers in that market at $3.00 each. The survey will attempt to evaluate market possibilities, determine consumer response, and iron out any possible kinks in the product as adapted for civilian use. Westinghouse is also reported to be supplying 5,000 of these containers to Gulf Oil Company for a similar test campaign in another southern city to be selected by Gulf. . . .

Practically every company interested in this field seems to be very close-mouthed as to its plans, and no publicity is wanted. In many cases their plans are not yet definite, and they hesitate to make any tentative announcement that might later have to be changed. The sudden end to the war apparently caught the insecticide industry unprepared, at least on aerosols, and many firms are simply not able as yet to say with assurance what their plans may be. Others are in the position, perhaps, of working on types of containers which they do not want their competitors to find out about, at least until they are past the test stage and ready to go on the market. . . .

Much speculation is noted as to the potential volume of aerosol sales. In the final analysis, this will probably depend to a great extent on the price at which an effective product can be sold. At four dollars—or even at two dollars—aerosols can tap only a small segment of the insecticide consuming market. There are always a small number of potential buyers in any market for a novel gadget such as the aerosol bomb, regardless of price, but to sell over a period of years, and in competition with oil base insecticides and the conventional type sprayers, it seems quite apparent that aerosols will have to come down to a lower price level.

Opinion in the trade indicates that the present cost of production of a one-pint aerosol bomb of the type used by the Army and Navy is in the neighborhood of $1.25 to $1.40. Perhaps production economics, use of a lower percentage of active ingredients, lower cost toxics, perhaps substitution of a cheaper gas for "Freon," may in the next year or two push this figure below a dollar, but even if this result is achieved, it is difficult to envision a low enough price to get many bombs of this type into the hands of the lower income brackets of the population, who incidentally are the largest buyers of insecticides as market studies indicate.

7

Achievements and Possibilities in Pest Eradication

CLAY LYLE

The recent progress in the development of new insecticides and insect repellents has not been equaled in all history. The discovery of the insecticidal value of DDT, closely followed by the amazing emergence of benzene hexachloride from a century of obscurity, has so stimulated research in insect physiology and in the chemistry and toxicology of insecticides as to make certain the synthesis of compounds even more remarkable than those now attracting the attention of the world.

The improvement in methods of application of insecticides merits no less praise. The brief time available does not permit listing all the machines for applying concentrated sprays and dusts by airplane and ground machines. Suffice it to say that at no previous time in history have the achievements of entomologists, working in collaboration with chemists and engineers, been of such universal value as to make in so short a time the name of an insecticide a common word in every household however humble or remote. The entomologist has become a wizard in the eyes of the uninitiated—and indeed some of the achievements seem little short of magic.

Unfailing evidence of this rise of the entomologist in popular favor is shown in increased expenditures for pest control. These increases in most

Journal of Economic Entomology 40 (February 1947), 1–8.

cases are far beyond those justified by rising prices and apparently indicate a willingness to follow the leadership of entomologists in attacking problems which have long needed attention.

With all of this scientific progress and with the world believing in our ability to accomplish great things, should we not consider whether our post-war plans in entomology—local, national and international—are as comprehensive and as challenging as this favorable situation justifies? Is not this an auspicious time for entomologists to launch determined campaigns for the complete extermination of some of the pests which have plagued man through the ages?

Certainly we should be encouraged by the achievements of those who worked without these modern weapons but had no lack of enthusiasm and courage for their tasks. We do not know who first thought of the possibility of eradicating a pest. Undoubtedly many of the early entomologists indulged in wishful thinking but did not have the insecticides or equipment to justify serious consideration of the problem. . . .

Eradication records in the United States—*Gypsy Moth.*—Apparently the first official extermination campaign began in Massachusetts in 1890 with the appropriation of $50,000 for work against the gypsy moth, *Porthetria dispar* (L.). Unfortunately this program was abandoned by the Massachusetts Legislature in 1900 because such good results were secured there was no longer any popular interest in control. . . .

Cattle Tick.—Almost immediately after the epoch-making discovery of Smith and Kilbourne,[1] the possibility of eradicating the cattle tick, *Boophilus annulatus* Say, received consideration. Apparently the first recorded statement is that of Dr. Cooper Curtice, who in 1896 said:

> I look most eagerly for the cleansing of even a certain portion of the infested territory under the direct intention of man, for it opens the way to pushing the ticks back to the Spanish isles and Mexico, and liberating cattle from disease and pests and the farmer from untold money losses. Let your war cry be: Death to the ticks.

This work began in North Carolina in 1899 with rather discouraging progress at first, for 7 years later only 12 counties had been released from quarantine. Only within the past two or three years has the continental

1. Smith and Kilbourne discovered that ticks carried Texas cattle fever, demonstrating an animal disease could be spread by an insect.

United States become practically free from this pest. The long story of discouraging political setbacks, the dynamiting of vats and the killing of inspectors, emphasizes the need for an outstanding educational program in advance of any compulsory undertaking of this kind. . . .

Argentine Ant.—The first successful effort at eradicating the Argentine ant, *Iridomyrmex humilis* Mayr., was at Fayette, Mississippi, beginning in 1922. The town was declared free of the ants in 1924. More than thirty other infestations of this ant have since been eradicated in Mississippi. . . .

Mosquito Eradication.—An outstanding instance of mosquito eradication is the elimination of *Anopheles gambiae* Giles from Brazil. This species was found in Brazil in 1930 and serious outbreaks of malaria occurred in a small area that year and in 1931. The Brazilian Ministry of Health, cooperating with the Rockefeller Foundation, concluded a successful eradication program by the end of 1940.

Apparently a very successful campaign is also being waged in Brazil for the complete eradication of the yellow-fever mosquito, *Aëdes aegypti* (L.). A very intensive program has been under way for several years and many large areas now seem to be free of this species.

In 1943 *Anopheles gambiae* was reported to have caused more than 130,000 deaths in Egypt. . . .

Insects Which Should Be Eradicated.—In considering the insects which should be eradicated in the near future, I am omitting, with one exception, those which are already subjects of federal eradication or control projects. The exception is the gypsy moth and I mention it only because of the greatly improved possibilities for eradication through the use of airplanes rapidly and inexpensively covering large areas with small amounts of DDT instead of the slow, costly and laborious application of lead arsenate spray used until the past year or two. Practically perfect control is being secured. R. A. Sheals, in charge of this federal project, writes:

> Certainly very significant progress has been made in gypsy moth control since the advent of DDT and the development of new and more effective types of distributing apparatus. During the year ending June 30, 1946, this Division and its cooperators treated in excess of 127 square miles of gypsy moth infested woodlands. This is more acreage than has been treated for gypsy moth control in any previous six fiscal years. Treatment by aircraft, including all expenditures incident to that type of work, plus ten per cent for overhead, cost $1.45 per acre. This may be compared with the cost of treating areas in previous years with arsenate

> of lead dispensed from high-powered ground spray equipment at a cost ranging from 15 to 25 dollars per acre. The effectiveness of DDT compared with arsenate of lead or other chemicals used is amazing. No persisting infestation has been found on any area treated by aircraft during 1944 and 1945.

Certainly there seems to be every reason for optimism about the success of this project in the not too distant future.

The House Fly and the Horn Fly.—The time has now arrived for the eradication of the house fly, *Musca domestica* L., and with it the horn fly, *Siphona irritans* (L.). These two should be considered together for in all rural and suburban areas the eradication measures would be used against both at the same time. The eradication of the house fly and the horn fly from the United States today would be a far simpler task than the eradication of the Mediterranean fruit fly from Florida in 1929. The program should be voluntary at first, with as much publicity as possible about the results secured and the feasibility of complete eradication. Public sentiment would then support any compulsory measures necessary for completing eradication. Already results in several sections of the country for the first year after the release of DDT for civilian use give promise of what may be expected when more sprayers become available and the spraying of livestock and barns is a general practice. Ray Cuff of the National Livestock Loss Prevention Board reports that half the cattle in Oklahoma and Kansas were sprayed this summer and quotes a Kansan as saying: "This is the first summer when we did not actually need screen doors." At State College, Mississippi, where cattle and barns were sprayed for horn flies and house flies, and garages, basements and various other locations were sprayed for mosquitoes, the scarcity of house flies was noticed by everyone, in spite of the most favorable fly-breeding summer in years. Some dairymen in Mississippi reported no flies in their barns for as long as six weeks at a time. In Oklahoma, the State Board of Agriculture secured 190 large army decontamination sprayers on trucks and sent them practically all over the state. . . . Last spring Idaho, through the leadership of W. E. Shull and H. C. Manis, was apparently the first state to promote a fly eradication campaign. A very effective poster carrying the slogan "No Flies in Idaho" was scheduled for display in 10,000 public places. Dr. Manis writes:

> During August and September when our fly populations are normally at their peaks, it was almost impossible to find a fly anywhere. . . . One of our major stu-

dents in entomology . . . set up two fly traps in the city of Moscow—one beside a slaughter house and another across town by some garbage cans—and during a two-week period was unable to trap a single house fly. He finally located 12 flies in a farm house and from those was able to start a laboratory culture. This could probably have occurred in any town in Idaho and on almost any farm. The results on a state-wide basis were so outstanding that it was almost beyond belief.

If the campaigns are pushed vigorously early next spring there should be many situations before the end of 1947 where a city or county health officer will be asking anybody who has seen a fly to report it, after which a spraying crew will visit the property to locate and destroy breeding sources. This is not a fantastic dream but is something that is almost certain to happen, and we as entomologists should be the driving force of the movement, enlisting the aid of health and veterinary departments, farm organizations, livestock associations, civic clubs and other agencies. If entomologists do not take the leadership in this program somebody else will. . . .

Screwworm.—The screwworm, *Callitraga americana* (C. & P.), does not over-winter north of Florida and the extreme southern part of Texas. From these areas it spreads north each season. Eradication from Florida during the winter would free all the southeastern states from attacks. Eradication seems quite possible technically, but will require the destruction of wild hogs and probably some other wild animals, in addition to the cooperation of all livestock owners. Cooperative work with Mexico may also make it possible to prevent the northward spread through Texas each year. It is a problem which challenges the best cooperative efforts of entomologists and livestock owners.

Argentine Ant.—This important pest has become established in the southern states and in California during the past 50 years. Although the present infestations probably comprise much less than one-tenth of 1 percent of the whole area of the infested states, its continued spread and eventual overrunning of all the South is almost certain unless control or eradication measures are begun. In towns and thickly-settled rural areas eradication is very practical and can be financed with local funds, as has been demonstrated at several dozen separate localities in Mississippi. Where large infestations occur in thinly-settled rural sections, the technical problem of eradication is no greater, but local funds must be supplemented from other sources. The eradication of the Argentine ant would pave the way for the eradication of the sugarcane mealybug and perhaps other species flourishing in a symbiotic relationship with it. . . .

I have not included widely-distributed pests which could be eradicated by eliminating the hosts, because of the difficulty of providing a satisfactory governmental basis for enforcing eradication. For example, the boll weevil, *Anthonomus grandis* Boh., could be eradicated within five years without much disruption of total production by means of a succession of non-cotton zones gradually extending across the cotton states, but no state would voluntarily adopt the legislation necessary to start such a program. There are probably several other pests, the eradication of which would be technically quite feasible but politically impossible.

I have excluded from present consideration most field, orchard, garden and forest pests, largely because our knowledge of the insecticides effective against some of them is not yet sufficient to make certain that we would not greatly disturb the "balance of nature" by large-scale field applications. This paragraph from a recent letter indicates that it is not difficult to create a faunal desert in some cases:

> Following five per cent gamma in talc, applied by airplane at rates of ten to twelve pounds per acre, the following were taken from the ground dead or paralyzed: aphids, beetles of several species, bollworm moths, bollworm larvae, leafworm moths, leafworm larvae, spiders, bees, *Geocoris*, aphis lions, flies of several species, tiger beetles, rapid plant bugs, tarnished plant bugs, *Neurocolpus*, velvetbean caterpillar moths, twelve-spotted cucumber beetles, grasshoppers, boll weevil parasites, ladybird beetles, leafhoppers and ants. In a field treated with twelve pounds per acre of the same material, applied 6 times by airplane, no insects were ever seen, although there were a few red spiders on practically every leaf.

Most of our economic plants are so dependent on insect assistance in pollination we cannot safely plan wide-scale distribution of such deadly materials until we know more about their effects on beneficial species.

In conclusion, may I plead for your serious consideration of the proposals for the eradication of these age-old pests. Let us not be satisfied with anything less than a post-war program which will challenge the imagination of the world. . . .

We have the technical knowledge and equipment to eradicate the house fly, horn fly, cattle grubs, cattle lice and several other insects, but in each case we must secure a high degree of control through voluntary participation before public sentiment will compel final eradication by law. Unless we can enlist the aid of farmers' organizations, public health agencies, schools, chambers of commerce, the press, civic clubs, city and county officials, legislators and members of Congress, we shall not succeed. . . . More than once during the past year I have heard the complaint that entomologists are not being accorded the recognition rightfully due them. Could it be possible that part of the fault may lie in us? Let us examine our programs carefully. In the words of Daniel Hudson Burnham, let us

> Make no little plans. They have no magic to stir men's blood.

8

The Food and Drug Administration Looks at Insecticides

PAUL B. DUNBAR

The Insecticide Industry Is Reminded of the Need for Recognizing that Discretion and Discrimination Must be Employed in Using Poisons

In the briefest and least technical way I want to tell you of the thinking of the Food and Drug Administration about the use and abuse of poisonous sprays in the production of foods. Our concern, of course, is from the standpoint of our obligation to protect the consumer through the enforcement of the Food, Drug, and Cosmetic Act.

FUNDAMENTALS

To begin with, let me state a few fundamentals. (1) The Food and Drug Administration recognizes that the use of insecticides is necessary both to bring many agricultural food crops to maturity in a condition suitable for human consumption and to protect many foods against insect depredations during manufacturing operations and storage. (2) By and large insecticides are poisons. If they were not poisonous they would of no value as insecticides. Their toxicity varies only in degree. (3) The terms of the Federal Food, Drug, and Cosmetic Act do not preclude the use of insecticides, but they do make provisions to guarantee that when they must be used consumer safety shall be assured.

Food Drug Cosmetic Law Journal 4 (June 1949), 233–39.

CONGRESSIONAL INTENT

In drafting this law, Congress obviously had in mind that under modern conditions we humans are exposed to traces of toxic substances from many sources. They recognize that the sum total of our intake of these minute quantities of toxic substances may be hazardous unless appropriate steps are taken to safeguard the public in every possible way. And so, when they came to legislate about the purity of foods they held that a food is adulterated if it bears or contains *any* poisonous or deleterious substance which may render it injurious to health, regardless of whether that substance is a natural component of the food or is added. They went further and said that any poisonous or deleterious substance *added* to a food shall be deemed to be unsafe regardless of the amount added, but, recognizing that this would outlaw the use of necessary insecticides and severely curtail food production, they provided an exception when it can be shown that the use of a poisonous substance is required in production or cannot be avoided in good manufacturing practice. In such cases they directed the Administrator to promulgate regulations setting safe tolerances with the force and effect of law. They enjoined him when setting such tolerances to take into account, not only the extent to which the use of the poison is required or cannot be avoided, but also the other ways in which consumers are exposed to the same or other poisonous substances.

That provision of the law imposed a heavy responsibility on the Food and Drug Administration. I confess that we have not fully met our obligation. The law was passed in 1938; the war intervened and made it impossible for us to hold the necessary hearings to establish tolerances. . . . Then we were confronted with a difficult situation. During the war a large number of new and very potent insecticides had been developed. Little was known about their toxicity either to the person who applied the sprays or to the consumer who ate the finished food product. In several cases we didn't even have methods for accurate estimation of the residual spray left on or absorbed by the food product. We didn't know whether the residues remained intact, whether they were altered by weathering to nontoxic or more toxic residues, whether they could be removed by washing, or whether they were absorbed into the plant structures and, therefore, could not be removed. We knew too little about many of these insecticides to hold hearings and establish safe tolerances.

PROBLEM OF CUMULATIVE TOXICITY OF INSECTICIDE

When the toxicologist talks about poisons he views the subject in two ways: first, the possibility of acute poisoning and, second, the possibility of chronic poisoning. Acute poisoning is something that doesn't worry us very much today in connection with spray residues on fruits and vegetables. Such poisoning occurs only through negligence. What the toxicologist *is* worried about is the long-time consumption of minute amounts of a poison which may eventually build up in the system to produce a serious physical disturbance. That is what they call cumulative poisoning. It has long been known that arsenic and lead have cumulative effects. We are just beginning to acquire knowledge about the cumulative toxicity of a few of the newer insecticides.

To avoid overtaxing your patience, I am going to talk from here on about DDT. It will illustrate the general problem of all insecticides. DDT is a tremendously useful pesticide. You began manufacturing it in enormous quantities about the middle of the second world war, and I don't think there is any doubt but that its use saved the lives of many thousands of our boys who otherwise would have succumbed to typhus and malaria. There wasn't any question even in those war years but that DDT was poisonous. Work in the Public Health Service laboratories and in our own, done at the request of the military authorities, proved that; but we were able to reach the conclusion that it was a reasonably calculated military risk to use DDT as a typhus and malaria preventative and that the risk of poisoning was less serious than the risk of exposure to these diseases. In reaching that conclusion the toxicologists believed that, while exposure to DDT would be fairly heavy, it would not be long-continued, and that the risks of cumulative poisoning would therefore be low. Since the war, DDT has deservedly retained its popularity as an extremely efficient insecticide. But, by the same token, there is an increasing danger of exposure of the general public to small continued intakes of DDT from many sources and for long periods with the resultant hazards of cumulative effect, particularly since the public has come to believe that it is not poisonous. We do not know how serious this hazard is in terms of human damage. We do know this: that experimental rats fed rations containing very small amounts of DDT, amounts of but one part per million or thereabouts will, in the course of time, and well before the end of a normal rat's lifetime, store DDT in the fat and with five parts per million will develop liver damage that is minimal but characteristic;

that female dogs exposed to cumulative effects of DDT secrete DDT in the milk; further, that mother rats fed fifty parts per million or more in their diet produce smaller weaning rats and fewer survivals in a litter than control animals. Now we cannot perform similar experiments on babies. We cannot exactly translate the results of rat experiments into terms of human effects. Certainly in any study of possible toxicity, if we know that one species of animal is affected, there is only one course to follow and that is to play safe and assume that the same results would follow if we could use human animals as experimental subjects.

NEED FOR DISCRETION IN USE OF POISONS

While DDT is an extremely useful insecticide, I am satisfied, as I have said, that the public has been more or less educated to believe that it is completely harmless. That undoubtedly has led to careless use by consumers in the household and by food producers. The other day I received a pathetic letter from a lady in Georgia who stated that she had a small house and garden in a country area, and that nearby farms were being literally deluged with poisonous sprays. To quote her: "Our home is in a smother of poisons from surrounding farms from early spring until late autumn. Our continued illness, particularly my own during this time, has occurred after being exposed to these poison mists and sprays." She goes on for several pages to recite the ills which she and her family have suffered, and she sincerely believes, and perhaps correctly, that these ills are attributable, at least in part, to these continued exposures. Apparently her doctor believes so too.

Now, gentlemen, as I said at the outset, poisonous sprays are necessary and a very valuable part of our program of producing sound and edible food supplies, and of combating harmful and destructive insect pests. But the time has come when I am certain that this industry which you represent should and does recognize that discretion and discrimination must be employed in the use of poisons and that the ever-present objective must be the protection of human beings from undue exposure.

DDT AND THE DAIRY INDUSTRY

Until a few months ago we didn't have the information we now have about the cumulative effect of small doses of DDT. Such experiments require several years. Only recently have the experiments of the Bureau of Entomology and Plant Quarantine been carried to a point where it was necessary for them to say that if dairy cows are fed silage bearing DDT, if dairy cows are sprayed with DDT, or even if DDT is sprayed in dairy barns, the milk from those cattle will contain DDT. Prior to that discovery, it was our view that, with proper precautions to protect the food from contamination, DDT was an effective and safe insecticide to use in connection with food production of every type. However, when inquiry was made of us, following the report of these results by the Bureau of Entomology and Plant Quarantine, as to whether DDT was a safe substance for use in dairy cow products, there was only one answer to make. Milk is a most important and universal food. It is the principal food of many babies from almost the day of their birth. It is an important food of children, and is important as an item of the diet throughout our lifetime. The Food and Drug Administration will not and cannot set up a tolerance for DDT in milk. It is plain common sense that dairy practice shall be so conducted as to protect the milk supply. Fortunately, the Bureau of Entomology and Plant Quarantine was ready to suggest alternatives and far less objectionable substitutes for DDT. I cannot say too much in praise of the action of the Bureau in promptly adjusting its spray recommendations to fit the new situation. There is no ground for hysteria about our milk supply. The dairy industry will certainly abandon its practice of using DDT in favor of some less objectionable substance. After all, in many areas of the country the fly population is no problem during the winter months, and in summer the cattle are in pasture, unexposed to DDT. The impression that a vast number of our population are consuming harmful quantities of DDT in milk is unfounded. Our spot checks of market milk throughout the United States justify that statement.

But the risk exists; and therefore I can do nothing but to advise you to back up your Executive Secretary, Mr. Hitchner, in his effort to encourage the *discriminating* use of insecticides.

SETTING UP SAFE TOLERANCES OF INSECTICIDES

A paper entitled "Pharmacological Considerations of Insecticides" was delivered by Dr. Arnold J. Lehman, Chief of our Division of Pharmacology, at the San Francisco meeting of the American Chemical Society in April. . . . I . . . want to mention his views about DDT. He says that in his judgment the danger point should be under one part per million if a large part of the food consumed is contaminated but that five parts per million approach that point if DDT is found only in single items. By that he means that if it can be shown, for example, that the spray schedules can be so controlled by the entomologist and by you manufacturers that the harvested crop of most fruits and vegetables will contain no DDT, but that, perhaps, apples and pears may contain some unavoidable residues, then it may be permissible to set up a legal limit of five parts per million for apples and pears. On the other hand, if it turns out that a great many other foods, in order to be brought to maturity without being destroyed or badly damaged by insect infestation, must contain some DDT, then a wide range of products must be provided with tolerances. In that case, since the consumer intake would come from so many sources, a tolerance as low as one part per million may be required. The Food and Drug Administration will never be able to, nor should it, set up a DDT tolerance for *every* variety of food product. Certainly, milk is one product where we will not do so. In general, baby foods of every type are in this category. Other foods which are less continually and less universally used may properly be permitted to be sold under a tolerance which quite possibly may not be as large as five parts per million but should be sufficiently liberal to permit effective but discriminating spraying.

I am quite confident that, with the determination on the part of insecticide manufacturers so to control their labels and distribution as to protect the consumer, and with the able guidance of the entomologists in the Department of Agriculture and in the various state organizations, it will be possible to work out a program that not only will protect the food supply, but also will guarantee consumer protection. Such a program should call for avoidance of the use of insecticides in food production unless a real need for such use exists. It should obviously envisage the abandonment for food production of any type of insecticide which is too poisonous for safe use on foods.

OBLIGATIONS OF THE INSECTICIDE MANUFACTURER

May I close with the suggestion that there is a moral obligation upon each of you to bend every effort toward the synthesis, and economical manufacture, of pesticides which show promise of combining low human toxicity with high insecticidal value. A corollary obligation, it seems to me, is research in your own organization on methods of chemical analysis of new pesticides that will afford an accurate index of residues not only on, but *in* our foods, in all those cases where presence of that particular pesticide is necessary. Such methods also will be invaluable to the industry and government pharmacologists in estimating the effects of the pesticide on human beings. There are heartening indications that certain of you are fully conscious of both these obligations and are conducting research which, while in the short view may not seem to be productive in profits, will, in my estimation, pay tangible dividends in the long run. I heartily congratulate those concerns and, to the others, say "Go thou and do likewise."

9

DDT and Its Effect on Fish and Wildlife

CLARENCE COTTAM AND ELMER HIGGINS

From the beginning of its wartime use the potency of DDT has been the cause of both enthusiasm and grave concern. Some have come to consider it a cure-all for pest insects; others are alarmed because of its potential harm. The experienced control worker realizes that DDT, like every other effective insecticide or rodenticide, is really a two-edged sword: the more potent the poison, the more damage it may be capable of doing. Most organic and mineral poisons are specific to a degree; they do not strike the innumerable animal and plant species with equal effectiveness: if these poisons did, the advantage of control of undesirable species would be more than offset by the detriment to desirable and beneficial forms. DDT is no exception to this rule. Certainly, such an effective poison will destroy some beneficial insects, fish, and wildlife.

Many exploratory investigations were made in 1943 and 1944 . . . [and the] need for more detailed information about the effects of DDT led in 1945 to concurrent studies to determine the effects of this poison upon insects, birds, mammals, fishes, and other wildlife. . . .

This year's work indicates that much remains to be done before the long-time effects of DDT on wildlife can be properly evaluated. . . . Further, studies are now under way to determine the effects of DDT upon marsh and

Journal of Economic Entomology 39 (February 1946), 44–52.

aquatic organisms, especially on the relation of malarial control to wildlife. Though many more studies are needed in all the above-mentioned fields, it seems that the most pressing requirement for the future is a study to determine the effects of DDT as applied to agricultural crops, to determine the effects of such control practices and procedures upon the wildlife and game dependent upon an agricultural environment. We must remember that about 80 percent of our game—and a very high percentage of our non-game and insectivorous birds and mammals—are largely dependent upon an agricultural environment. The heaviest and most widespread application of DDT will probably be used in this environment; therefore, it is not improbable that the greatest damage to wildlife will result here. As has been demonstrated in the field of forest-insect control, a well-coordinated study in the application of DDT to agricultural crops will greatly minimize this damage.

[The authors describe several experiments, omitted here.]

Lackawanna County, PA.—A breeding-bird census similar to that at the Patuxent Refuge was taken on three forty-acre areas of forest land near Scranton, Pennsylvania, between May 1 and June 27, 1945.

Between May 24 and June 1 the experimental control of gypsy moth larvae was begun by the application (by airplane) of an oil spray of DDT at the rate of five pounds to the acre on a 600- acre tract enclosing the first of these three bird-censused areas. The gypsy moth and other tree-feeding caterpillars were eliminated and many other insects were affected within a few hours. Some species were apparently unaffected. A hive of bees placed in the tract before it was sprayed came through in good condition. Several groups of insects that were reduced by DDT had again attained their normal numbers within three months.

Within forty-eight hours after the DDT was applied, five birds showing symptoms of DDT poisoning were found; all died. In addition, two other sick birds were found and two nests were abandoned. Species killed included red-eyed vireo, black and white warbler, ovenbird, redstart, and scarlet tanager. Within forty-eight hours after the application of DDT to the final portion of the area on June 1, the population of living birds (which had been 1.6 pairs to the acre before spraying) was much reduced. On June 13 the area the area contained only 0.5 birds an acre. The most commonly noted species were ovenbird and red-eyed vireo.

On June 9 an oil spray at the rate of one pound to the acre was applied by airplane to a 350-acre tract enclosing the second forty-acre bird-censused area. A nearly complete kill of gypsy-moth larvae and a conspicuous reduction in the general insect population were effected, but at no time were insects scarce. Most groups apparently recovered within a month.

The third area remained untreated as a control for the second.

Before spraying, the population of all birds in these areas was 2.7 pairs an acre. The commonest species were ovenbird, red-eyed vireo, Canada warbler, magnolia warbler, northern water-thrush, and redstart. After spraying, the population was reduced to 2.6 pairs to the acre in the sprayed area and 2.4 in the control area, owing, presumably, to completion of nesting or to destruction by predators.

Fishes.—On August 9 a three-mile section of Ash Creek, near Clifton, Pennsylvania, was sprayed by airplane with an oil solution of DDT, applied at the rate of one pound an acre as the plane followed the course of the stream. . . .

A census taken before spraying indicated a total brook-trout population of 2100 in the sprayed section. There were also many common suckers, common shiners, black-nosed dace, fallfish, and golden shiners, particularly in the lower mile. . . .

Poisoned fish began drifting into the weirs twelve hours after spraying. Sixty-nine percent of all observed fish losses occurred within thirty-four hours after spraying.

The warmer-water fishes—common shiners, common suckers, and golden shiners—were affected first, most of their mortality occurring within two days after spraying. It was estimated that less than half of them were killed.

Brook and brown trout were more slowly affected; their losses, as well as the loss of common suckers, lasted for a week. Sampling of the native brook-trout population before and after spraying indicated that only about 1.3 percent of the brook-trout population (twenty-seven fish) were killed.

In the live boxes stocked with brook trout, the greatest loss was four fish, at the weir farthest downstream, in live boxes stocked with miscellaneous cold-water fishes—including sculpins, black-nosed dace, creek chubs, pearl minnows, top minnows, and common suckers—only four fish were killed out of a total of 203. . . .

Island Beach, Ocean County, New Jersey.— To eliminate mosquitoes, the southern five miles of Island Beach—including the shallow shore waters in Barnegat Bay—were sprayed by airplane on July 11 with an estimated 0.5 pound of DDT to the acre. Partial control was achieved.

On July 8, previous to spraying, large numbers of gulls, terns, herons, shorebirds, barn swallows, purple martins, redwings, and other song birds were observed. No direct effects on the birds were noted after the spraying, but most swallows and other insectivorous species left the area and remained away the rest of the summer. No sick or dead birds were found.

Blue crabs, plentiful along the bay shore before the spraying, moved out of the sprayed shore waters after the spraying but were abundant in unsprayed areas. No dead crabs were found three days after the spraying. By July 18 reports were received of the death of many blue crabs that had reappeared in the sprayed area; on July 21, 150 dead or dying crabs were counted on a 200-yard stretch, though crabs in adjacent unsprayed waters were healthy. On the latter date many active fiddler crabs and ground-inhabiting insects were in evidence and no dead birds were observed.

Along five miles of the shore on Barnegat Bay there were an estimated 100,000 small dead fishes (menhaden, mullet, and killifish) upon which tern and gulls were busily feeding. . . .

Other field observations.—DDT was used on an experimental basis for insect control in several other localities, but on these the Fish and Wildlife Service personnel involved could not make detailed wildlife population studies. Careful qualitative observations were made, however, and in some of the localities damage by DDT was noted.

Summary.—1. Observations on the damage to wildlife caused by DDT used in insect control were made in twelve states and the Province of Ontario. Fair to successful insect control was effected.

2. The principal investigations were made on forest lands in Maryland and Pennsylvania, where forest tree pests were the objects of control.

3. Application of DDT was made chiefly by airplane as an oil spray. Spray concentrations ranged from 0.2 pound to five pounds an acre. The greatest amount used anywhere was twenty-five pounds an acre in the form of a dust.

4. The amount of DDT that actually reached the vegetation showed considerable local variation from the specified rate of application, owing to bad weather, lack of landmarks, defective spray apparatus, and different density of vegetation.

5. Pronounced mortality in wildlife resulted from the use of most of the higher concentrations of DDT; mortality was slight or not apparent in most of the lower. Invertebrates and cold-blooded vertebrates were more readily affected than were birds and mammals.

6. In the single trial involved, DDT was much more toxic in emulsion than when applied in oil or as a suspension.

Recommendations.—1. *Don't Use DDT Unless You Must.* Undertake its use for the control of an insect pest only after weighing the value of such control against the harm that will be done to beneficial forms of life. Wherever more than a small area is involved consult county agricultural agents, State or Federal entomologists, wildlife and fishery biologists, and U.S. Public Health Service officials.

2. *Always Use Minimum Necessary Dosage.* Use less than 0.5 pound of DDT an acre in an oil solution to avoid damage to fishes, crabs, or crayfish; less than two pounds an acre to avoid damage to birds, amphibians, and mammals in forest areas. Do not use emulsions.

3. *Apply Carefully.* Use DDT only in calm, dry weather and only where it is needed. Wherever it is applied by airplane, provide careful plane-to-ground control to insure even coverage and to prevent local overdosage.

4. *Leave a Sanctuary.* In forest pest control, alternate treated and untreated strips in order to disturb wildlife as little as possible.

5. *Time the Application.* Making use of its residual effect in the control of early appearing insect pests, apply DDT just before the emergence of leaves and the main spring migration of birds; for late-appearing pests, delay applications if possible past the nesting period of birds (to mid-August). Adjust crop applications and mosquito-control applications as much as possible to avoid the nesting period.

6. *Avoid Spraying Water.* Because of the sensitivity of fishes and crabs to DDT, do not apply directly to streams, lakes, and coastal bays.

7. *Watch What Happens.* Wherever DDT is used, make careful before-and-after observation on mammals, birds, fishes and other wildlife.

PART 3

RISING CONCERN ABOUT NEW PROBLEMS

This part traces the growing knowledge of DDT's effects—mainly through scientific studies on wildlife, for that remained the center of interest until *Silent Spring*—and growing concern. The first paper, by an ornithologist, Roy J. Barker, records an early encounter with two of DDT's most alarming properties, its persistence and bioconcentration (rising levels of residues at higher levels of food chains) in a very local setting. The editorial from *Bird Study*, the scientific journal of the British Trust for Ornithology, and the paper from the same issue by the ornithologist Derek A. Ratcliffe, "The Status of the Peregrine in Great Britain," suggest that what Barker found on the University of Illinois campus might be happening across Britain. The next, by Robert Rudd, an ecologist, presents some important cases that led scientists to oppose widespread use of DDT. Joseph Hickey then tells how and why he changed his mind about DDT. Hickey had been a bird-crazy teenager in the Bronx when he met Roger Tory Peterson; he became an ornithologist due to the encouragement of Ernst Mayr, a German academic working for the Museum of Natural History in New York, whom he met on birding expeditions. His account shows how scientists' standards of proof differed from the public's, but it also traces the connections among researchers, as one paper stimulated another and suggested a third, conversations opened still other lines of thought, and pieces fit together to make a picture. While scientists traced residues through the environment, ordinary people worried about dead robins on their lawns. The last piece, Robert S. Strother's "Backfire in the War against Insects," gave an account for the public of the scientific evidence. That the article appeared as a staff piece in *Reader's Digest*, the very ideal of middle-class, middle American respectability, suggests how much concern had grown since the time, only a few years before, when parents watched their children run through the fog laid down by municipal trucks spraying beaches or suburban streets.

10

Notes on Some Ecological Effects of DDT Sprayed on Elms

ROY J. BARKER

The main purpose of this paper is to call attention to the possibility that moderate applications of DDT under certain conditions can be concentrated by earthworms to produce a lethal effect on robins nearly one year later.

When phloem necrosis and Dutch elm disease threatened American elms (*Ulmus americana*) at Urbana, Illinois, in 1949, DDT was sprayed in attempts to control suspected insect vectors. Subsequently, some moribund, tremoring birds were observed. This report concerns an exploratory investigation to determine the circumstances that caused the ensuing mortality in birds and other adverse effects.

INSECTICIDE APPLICATIONS

From 1949 to 1953, 1,400 American elm trees on the 430-acre main campus of the University of Illinois in Urbana were sprayed twice each year with a 25 percent DDT emulsion concentrate of twenty lbs. of technical DDT in five gal. of xylene, 2.5 gal. of Acme white oil, and 1.5 pts. of Triton X-100. . . .

Although elms were the predominant shade tree in the adjoining cities of Champaign and Urbana, the only obvious control efforts were by a few individual home owners who contracted with pest-control operators to

Journal of Wildlife Management 22 (July 1958), 269–74.

apply similar DDT sprays. The campus was essentially a sprayed plot surrounded on three sides by an unsprayed elm forest.

During application, DDT dripped or drifted directly to the soil. . . . The average deposit in this sampled elm grove was 1.7 lbs/a [pounds per acre]. . . . The results . . . indicate that the quantity of DDT in the soil does not decrease over winter but might actually increase with the fall of sprayed leaves. . . .

EFFECT ON BIRDS

Very few dead birds were observed in 1949 when DDT was first sprayed. However, before spraying began in the spring of 1950 the numbers of dying robins (*Turdus migratorius*) had attracted unsolicited attention. Dying birds were reported frequently after rains, and at that time a lethal dose of DDT was suspected of having been acquired by the birds drinking water from puddles in contaminated soil. Analyses of combined ether extracts of water from three or four puddles in areas that had been sprayed in 1949 indicated that DDT was present in the spring of 1950 only in trace quantities.

Most dying robins exhibited typical tremors associated with DDT poisoning. Because brains are easily dissected, these were analyzed to confirm the presence of DDT or DDE. The data . . . suggest that the DDT content of dying birds was affected only indirectly by DDT sprayed in mid-June or by seasonal variation in food. . . .

Several animals, other than robins, were found with apparent DDT poisoning. The brains of these were analyzed for DDE and DDT. Evidently one common grackle and two house sparrows had been able to accumulate a lethal dose of DDT. . . .

SUSPECTED ROLE OF EARTHWORMS

The collected robins had empty crops. This would suggest that they had acquired DDT in a material that was quickly digested. . . . Because dying robins were found in the spring before insects became abundant and because more robins died after heavy rains, earthworms were suspected of containing a lethal dose of DDT. . . .

Seven specimens of *Lumbricus terrestris* were collected on July 17, 1950, from various parts of the campus. These were dissected and the tissues analyzed to find if they contained DDT. . . . In view of the low concentrations found in the soil, [the analytical result] suggests that by selective feeding on sprayed leaf mulch, these worms were concentrating DDT. Since feeding habits could influence DDT uptake, several species of earthworms were analyzed for DDE and DDT. . . .

DISCUSSION AND CONCLUSIONS

A fifty-gram robin with sixty µg./g. [micrograms per gram] of DDT contains the DDT equivalent of eleven *L. terrestris* or about sixty of the smaller species of earthworms examined in this study. Since robins have been reported to take ten to twelve earthworms in as many minutes (Barrows, *op. cit.*), and since nestlings will each day consume food equal to 50 percent of their body weight in addition to that supplied by the parent birds (George and Mitchell, 1947), earthworms seem to be one likely source of DDT poisoning for this species. . . .

The immediate effects, or lack of any, caused directly from DDT sprays have been assessed by other investigators. Direct sprays of DDT on nests (Mitchell, 1946) did not reduce egg hatch. Sprayed spruce budworms or corn borers fed to nestlings (George and Mitchell, *op. cit.*) gave toxic responses.

Earthworms furnish an example of the complexities in assessing long-range effects of toxic residues. These biological concentrators of DDT are influenced by such factors as soil structure, cultural practices, and weather (Walton, *op. cit.*). The meager data available emphasize that biological effects of toxic chemicals or isotopes cannot be assessed in a few weeks following application.

11

Editorial from *Bird Study*

For many hundreds of years, man has influenced the distribution and abundance of birds in the British Isles. Many species have been exploited as sources of food or plumage, while changes in agricultural practice and the increase in the human population have radically altered the distribution of many habitats. Within the last hundred years, this influence has probably reached a level far greater than ever before with an improvement of fire-arms and means of transport. . . .

This number of *Bird Study* includes a detailed paper on the status of the peregrine. The Council of the B.T.O. [British Trust for Ornithology] have considered that the results of this enquiry are so important, and the implications so far reaching, that priority has been given to its publication. This study clearly suggests that "toxic chemicals" which are used in agriculture are the most important single factor responsible for the recent rapid decrease of this falcon, either killing individuals or producing the effects of infertility in individuals which have accumulated sub-lethal doses.

How many other species of birds are decreasing or have decreased like the peregrine? It is very important to have an answer to this question, but this cannot even be attempted with our present knowledge. When it is remembered that a relatively rare species such as the peregrine was not

Bird Study 10 (June 1963), 55.

detected as declining in the British Isles until this enquiry was carried out, and that this was instigated by the Nature Conservancy as a result of repeated representation by racing pigeon interests, who considered that the falcon had increased in numbers and was now too plentiful, it is clear that our knowledge of the quantitative status of most species is inadequate. Nevertheless, there can be little doubt that many species of birds are being affected by toxic chemicals and this is effectively undoing over a century's work of bird protection. More information must be obtained regarding bird population trends and also toxic chemicals. This must be done soon and the Nature Conservancy have established a research group to study the effects of toxic chemicals. This work will cost a considerable amount of money and it is to be hoped that the Government will be much more generous over grants for this work than it has been for other aspects of the work of this body. It should be remembered that the amount of time available for this work is limited; not even a government department can recreate a species which is extinct.

12

The Status of the Peregrine in Great Britain

DEREK A. RATCLIFFE

Between 1900 and 1939 the available records suggest that in many parts of Britain the peregrine population remained relatively constant, with only minor fluctuations in numbers. . . . The experience of a large number of egg collectors, falconers, gamekeepers and others who have followed particular territories over a long period is that, as a rule, a peregrine nesting place, or its alternatives, is occupied regularly, year after year, for an indefinite time. Many nesting cliffs have a long history of occupation, extending back well over a century, and a few were known to be occupied in Stuart and Tudor times. . . .

The war period 1939–45 brought a very drastic change in the peregrine situation. The species was outlawed as a killer of carrier pigeons, and the Air Ministry waged a campaign of control which resulted in its virtual extermination as a breeding species in the southern counties of England. . . .

The resilience of the species in the face of adversity was shown by the rapid recolonisation of many vacant territories in the shot-out districts when war-time persecution ceased. . . .

Yet, by the mid-1950s it was evident that several southern districts depleted by war-time persecution showed only a partial recovery to the pre-

Bird Study 10 (June 1963), 56–90.

war level of population. By the time the present enquiry was launched in 1961, there were clear signs of decrease in nesting populations in southern England, and very few young were being reared by the remaining pairs. In north Wales, northern England and southern Scotland low nesting success became prevalent and was associated with the frequent breakage or disappearance of eggs, and the evidence indicated that these were eaten or otherwise destroyed by the owners themselves.

The results of the enquiry show that in 1961 and 1962 the peregrine had again practically disappeared as a nesting species in southern England, despite the promising early post-war recovery. . . .

Even allowing for the case where territory desertion is suspected but not proved, the peregrine population of Great Britain has undoubtedly suffered a tremendous decline since the pre-war period 1930–39, and most of this decrease has occurred since 1955. This trend is the more remarkable in view of the potential for increase between 1945 and 1955, shown by the recovery in districts "controlled" during the war; even up to 1959 there were attempts (sometimes successful) in some districts to recolonise ancient nesting haunts which had been deserted for many years. . . .

It is interesting, though disturbing, to find that the peregrine is suffering a serious and synchronous decline in other parts of the northern hemisphere. Substantial decreases in breeding population are reported from France, Germany, Finland, Sweden and the eastern United States. A slight decrease is reported from Norway.

Decline on this scale surely indicates some extremely widespread influence and, of the possible "natural" factors, only climatic change would seem likely to affect all these countries simultaneously. . . . Yet, remembering that the peregrine is virtually a cosmopolitan species, occurring from the Arctic to the Tropics, the relatively slight amelioration of climate in the Palaearctic Region during the last half century (which some climatologists believe is now being reversed) is not likely to have had any significant effect. . . .

There remain four feasible explanations of decline, and only the first could involve "natural" environmental change.

Decrease in food supply on an unprecedented scale may be taking place. . . . While diminishing food supply appears to have caused the long-term decline in the peregrine population of the western Highlands and Islands, there is no reason yet to suppose that it could be responsible for the more recent and rapid decline in regions to the south. . . .

Shooting, trapping and poisoning of birds, and spoiling of nests undoubtedly still occur locally, but there is no evidence that such persecution is any worse than before 1939 (when it did not reduce breeding populations), or that it anywhere approaches the systematic destruction of the war years. In view of the powers of recovery shown by the species in the early post-war years, it would seem impossible for surreptitious persecution to be waged on a scale nearly sufficient to account for the rate and extent of the recent decline. Moreover, it is extremely unlikely that a similar measure of persecution is simultaneously affecting all the countries of the world where serious decrease in peregrines is reported.

The incidence of disease in the peregrine is unknown, and the few recorded cases of death by infection throw no light on the importance of this factor as a cause of mortality in the species. . . . In the complete absence of any positive evidence, such an explanation of the peregrine's decline cannot be advanced seriously.

Peregrines may be the secondary victims of agricultural toxic chemicals, through repeatedly taking prey which carried sub-lethal doses and so building up poison in the body, finally to a fatal concentration. The evidence for progressive build-up of pesticide residues along food chains has been collected by Carson in North America. These substances or their similarly persistent and harmful break-down products accumulate selectively in various organs and tissues of an animal or bird, and predators remotest from the source of the poison can build up high and lethal concentrations in their bodies, while organisms at lower levels in the food chain carry only sublethal doses.

In the United States, a disastrous fall in breeding success is threatening the existence of the bald eagle (*Haliaeetus leucocephalus*) outside Alaska, and is correlated with the presence of high concentrations of metabolites of DDT in the tissues and eggs of this species. Captive bald eagles fed on diets containing varying amounts of DDT all sickened and died, the severity of the symptoms and rapidity of death being proportional to the dose of insecticide. Likewise, an osprey (*Pandion haliaetus*) colony of some sixty pairs in Connecticut produced only a handful of young in 1961 and 1962, and addled eggs contained high concentrations of DDT and DDE.

In Britain . . . there is every reason to believe that the virtual disappearance of both kestrel and sparrow-hawk as breeding species in the south-east of England is connected with the copious use of pesticides in this region. . . .

The evidence that sub-lethal doses of chlorinated hydrocarbons cause first reduction in hatching success of eggs and then sterility of adult birds agrees with observations on the peregrine, in which an actual decline in numbers is preceded by a reduction in breeding success and by failure to attempt breeding. Moreover the frequent destruction of their eggs by peregrines themselves could be explained as abnormal and pathological behaviour due to the physiological upset produced by sub-lethal doses of poison.

. . .

The wave-like spread northwards of decline coincides closely with the pattern of use of organic pesticides, both geographically and in time. These substances first came into general use from about 1945 onwards, *DDT* being one of the first and most widely used. Since 1950 there has been an ever-growing use of newer, highly poisonous chemicals such as *Dieldrin*, *Aldrin*, *Endrin*, *Heptachlor* and *Lindane* (*gamma-BHC*). Such pesticides were used first and most extensively in the main arable farming and fruit growing districts of south-eastern England, the Home Counties and the Midlands, and these have remained the chief centres of use. But more recently, and particularly through the frequent dressing of seed by seedsmen, these chemicals have become used over the greater part of Britain where arable farming is practised. The degree of use is probably closely related to the proportion of land under plant crops other than grass. . . .

The decline of the peregrine started in southern England, and by 1956 there were signs that the post-war recovery had been halted or even reversed, leaving numbers below the 1930–39 level. Egg-eating, which I believe to be a first symptom of decline, was first noticed in 1951 and has been a common occurrence ever since. By 1958 a decline in breeding populations was well established in southern England and Wales. It has since accelerated there almost to extinction point and spread rapidly to northern districts. As has actually happened, sub-lethal effects would be expected to show first as the influence of toxic chemicals spread to new areas, but as contamination increased to the point at which death of adults occurred, real decline in the peregrine population would follow. . . .

The widespread use of pesticides in the eastern United States, Germany and France would account equally for the decrease of peregrines there. Toxic chemicals are not used on a large scale in Scandinavia, but the decline in Finland and Sweden might be explained by movement of peregrines during autumn and winter to more southerly parts of Europe, thus greatly increasing the chances of contamination.

The case against toxic chemicals was based on such circumstantial evidence at first, but one vital piece of direct evidence can now be given in support. On 1 July 1961 the writer visited a peregrine eyrie near Crieff, Perthshire, and found two addled eggs which the female was still incubating despite the lateness of the date. There were fragments of at least one more egg which had evidently been destroyed by the owner. The eggs were taken, and though one was later broken and lost, the other was sent to Dr. H. Egan in the Laboratory of the Government Chemist. Dr. Egan examined the egg by gas chromatography, and found that it contained 115 micrograms of *pp-DDE* (a metabolite of *DDT*), fifty micrograms of *Dieldrin*, twenty-eight micrograms of *Heptachlor epoxide*, not more than two micrograms of *Heptachlor* and not more than two micrograms of *gamma-BHC*. The total concentration of chlorinated hydrocarbon residues was of the order of four to five parts per million.

These findings, which have been reported in detail by Moore and Ratcliffe [in another article], prove that peregrines can take up and accumulate pesticide residues from their prey. It is extremely unlikely that the four different residues came from the same prey individual; more probably they were accumulated over a period from various items of prey. The potential vulnerability of the species is thus proved. . . .

Even though more chemical analyses of peregrine tissues or eggs are needed finally to clinch the matter . . . it is a reasonable conclusion that agricultural pesticides are by far the most likely cause of the recent and serious decline which has spread northwards through Britain.

The present state of the peregrine population thus matches the varying use of pesticides in different parts of Great Britain. The intensification and spread of the decline reflect the increasing contamination of the prey populations. Even at the present level of prey contamination, deterioration in the peregrine situation could continue, due to the time-lag in appearance of cumulative effects. Future prospects for the species are extremely bleak. This is a completely new factor in the biology of the peregrine, against which the bird appears to be defenceless, for there are no signs of any physiological adaptation to the toxins. At the present rate of decline, numbers and breeding success will soon be reduced to an extremely low level, and final extinction cannot be dismissed as a possibility.

13

Pesticides and the Living Landscape

ROBERT RUDD

The delayed expression of toxic symptoms among biological concentrators of pesticides is a serious problem primarily because it is insidious. It goes unnoticed until mortality results, and even then the cause of mortality may often not be suspected. . . . The illustrations below were barely hinted at twelve years ago, and only in the past few years has their significance been established. Delayed expression is a result of the general distribution of stable pesticides. I have no doubt that it is more profound and more widespread than the examples below indicate.

The greatest hazard accrues with repeated applications of stable chemicals. . . . [The] examples of delayed expression [given here] are not equally supported by fact, but ecologists will agree, I am sure, that the sequences are reasonable. Only small gaps in information remain to be filled; the evidence for delayed expression of transferred chemicals is no longer tenuous.

. . .

DELAYED EXPRESSION IN AQUATIC COMMUNITIES

Example 1.—The death of fish-eating birds following DDD applications to control gnats at Clear Lake, California.

Clear Lake is a shallow, warm body of water of about 46,000 acres, 100

Pesticides and the Living Landscape (Madison: University of Wisconsin Press, 1964), 248–67.

miles north of San Francisco. Its high productivity is reflected in excellent sport fishing, one of the attractions that has made Clear Lake an important tourist and recreation center. Unfortunately, insects also share this high productivity. The Clear Lake gnat (*Chaoborus astictopus*) particularly qualifies as a serious nuisance. Although it is not blood-sucking, its numbers during the warmer months are so great that it becomes an intolerable nuisance. Efforts to control the gnat have been made for many years, but it was not until 1948 that an effective method not hazardous to fish was found. The target zone was the bottom muck of the lake in which the gnat larvae develop. DDD—a near relative of DDT—was applied to the surface of the lake and mixed in as much as possible. A concentration of one part DDD in seventy million parts of water was sufficient to kill larvae. It also killed large numbers of other aquatic invertebrates, but did little immediate harm to fish populations. An application of DDD was made to the lake in September of each of the years 1949, 1954, and 1957. The second and third applications were at the higher rate of one part DDD to fifty million parts of water. . . .

Before the first application more than 1000 pairs of western grebes (*Aechmophorous occidentalis*) bred at the lake; next year the summer breeding colony did not return. A few pairs (fifteen to twenty) of birds were present during the breeding seasons of 1958–60, but no young were brought off. In 1961, sixteen nests of western grebes were found, but no young were fledged. One young bird was seen in 1962. Three young were observed in 1963. Grebes had continued, nonetheless, to visit the lake in large numbers during the winter. In December, 1954, death among these grebes was widespread; over 100 were known to have died. A later die-off occurred in March of the following year. In both instances, wildlife-disease biologists could find no evidence of disease. Following the third DDD application, in December, 1957, a third die-off was observed. Disease was also looked for on this occasion, and not found. As an afterthought, the visceral fat of two grebes was removed and analyzed for DDD. The astounding level of 1600 parts per million was found in these tissues—a concentration 80,000 times as great as that applied to the lake. Contaminated food was judged to be the source. This conclusion initiated the collections [of samples from the food chains in the lake] that were to make clear the sequence of biological concentrators, which perhaps constitutes thus far the best example of the delayed toxic expression of transferred chemicals. Collections of many kinds of fishes and other organisms have been made since 1958, and will continue until no trace of DDD can be found in tissues. Upon recommendations of the Lake County Mosquito Abatement District, no further applications of DDD to the lake have been made. . . .

The Clear Lake accounting, incomplete as it is, still stands as a fascinating example of the complexities of chemical transferral in living systems, and as a pointed warning to professional workers not to oversimplify their solutions to pest control problems. . . .

DELAYED EXPRESSION IN TERRESTRIAL COMMUNITIES

Example 1.—Mortality among birds many months after the application of DDT at two to five pounds per tree to control bark beetles.

Only a few years ago the accounting I am about to give would have seemed incredible. Now, both the fact of extensive losses of birds and the routes by which loss is caused are carefully documented. The only unknown features remaining are why pest control workers failed for so long to concede that loss was occurring and what the actual losses in bird numbers must be. The truth has finally penetrated control circles, but the total losses of birds cannot be known. Probably they range into millions. Among current control programs only the fire ant control campaign can have had effects as great.

The data now available are too voluminous to record here in detail. . . . Other work in progress will swell the documentation. The main emphasis here will be on the chain of chemical transferral leading to the delayed expression of toxic symptoms and mortality in birds. . . .

The fungus that causes Dutch elm disease was discovered in the United States only thirty years ago. Since the early recognition of the disease in Ohio and New York, it has swept through some twenty states, leaving behind millions of dead and dying elm trees. DDT gave promise of controlling the two species of bark beetles primarily responsible for the spread of the fungus. Dutch elm disease control programs were quickly put into effect by many communities, with some local success. Nonetheless, the affliction continued to spread. In 1949 it had reached Illinois; in 1950, Michigan; in 1956, Wisconsin; in 1957, Iowa.

[Rudd summarized the results from Barker's study and went on to say:] Precipitous declines among many species of birds have in fact been the rule where Dutch elm disease control with DDT has been repetitively practiced. Wallace *et al.* (1961) record a decline in numbers of robins on the main campus of Michigan State University, dropping in three years from 370 to fifteen to four birds. Nesting failures were more startling; virtually no broods were brought off during the years after spraying began. Hickey and Hunt (1960) estimate total robin mortality to be at least 85 percent for each spraying. In several Wisconsin communities consecutively sprayed for three years, total breeding-bird populations were reduced by 30 to 90 percent. Where there were as few as ten elm trees per acre and these were sprayed annually, the drop in population was about 90 percent.

Although robins seem to be the most obvious victims of DDT treatments for Dutch elm disease control, many other species are affected and there is no doubt of widespread declines in affected avifaunas. . . . The death of large hawks and owls cannot yet be understood. But, whatever the feeding group or the manner of poisoning, it can be convincingly demonstrated that bird populations have been widely reduced.

The question, as frequently posed, is not simply one of choice between the elm trees or the birds. Neither safeguarding trees nor safeguarding birds can be justified on strictly economic grounds, but both are valued for the comfort and pleasure they afford; both must be safeguarded. The search for acceptable, economical, and simple control methods must go on for the best interests of all. Unquestionably, DDT treatment has resulted in saving many elms and in slowing the spread of Dutch elm disease. But the treatment has also resulted in rapid declines in bird populations. Control purposes should never be defined so narrowly as to hazard general advantage, whatever sanctity can be assumed within the limits of single purpose.

[*Example 2*].—Mortality and population declines among vertebrates in the southern states on areas treated with heptachlor (primarily) at 1.5 lbs. per acre in efforts to eradicate the imported fire ant.

The fire ant control program has resulted in heavy immediate mortality to vertebrates, declines in population numbers, slow rates of recovery, and residual contamination of tissues with heptachlor and heptachlor epoxide in most of the fauna in the treated area. . . .

Earthworms on woodcock wintering grounds in Louisiana contained up to twenty p.p.m. of heptachlor epoxide in tissue six to ten months after land treatment. As just noted, woodcock reflect these concentrations by storing significant amounts of the epoxide in their own tissues. As yet, there is no indication of direct mortality from this source of contamination.

Carnivorous mammals contained residues that must have come largely from contaminated foods. Cotton rats, for example, which serve as an important food for carnivores, had a residue level of 5.9 p.p.m. heptachlor epoxide in tissue. Fishes and amphibians, which are fed upon regularly by vertebrates, almost without exception had detectable amounts of heptachlor epoxide. Reptiles too were commonly contaminated. In all these groups that serve as food, the source of tissue contamination must be their own food species. A centrarchid fish (bluegill), for example, with a residue level in tissue of 36.2 p.p.m. could have obtained such concentrations and still survived only if the epoxide was gradually accumulated from food sources. Crayfish with 1.6 p.p.m. in tissue after months of exposure could have acquired it only through ingestion of lesser amounts. One-tenth of that concentration in water would effectively obliterate entire populations of both bluegill and crayfish. Food-chain transferral of chemicals must be one of the commonest sources of effects following survival to the initial exposure. The final linkage of transferred chemicals and long-delayed effects has yet to be made, however (data from DeWitt and George, 1980).

14

Interview with Joseph J. Hickey

THOMAS R. DUNLAP

How did you get interested in the issue of DDT and come to see it as a hazard?

It started when I went to a funeral in Kenilworth, Illinois, in June 1958. There was a mulberry tree in front of the house. It was loaded with berries and there were no birds. I realized that something was extraordinarily wrong so I asked the local people "Where are the birds? Where are the robins?" They said there were robins earlier in the spring but they all went north. When we got out of town and to a golf course in the next town I saw robins. So I came back to Madison realizing that the reports from Audubon ladies in the Illinois Audubon Society were probably not exaggerated. I had initially dismissed them because they published their information in the journal of the Illinois Audubon Society, *not* a scientific journal but there was something very strange about this incident. And so that fall I put a student to work in the classical fashion in which I had some experience, and that is: what is the LD-50[1] for robins? We got an LD-50 . . . and we compared this LD-50 to that of the methoxychlor, the alternative chemical . . . and in comparison methoxychlor was a much more attractive alter-

Edited transcript of editor's interview with ornithologist Joseph J. Hickey, 19 July 1973. Tape of interview in Wisconsin State Historical Society.

1. LD-50 refers to a "lethal dosage" for 50 percent of the subjects receiving it.

native chemical. The following spring I had a small grant from the Department of the Interior to put a student to work on three communities in the state which were using DDT for elm disease control and three which were not—actually six study areas that were sprayed and five that were not. While the student was in the field, the campus of the University of Wisconsin in Madison was suddenly sprayed, without warning, and they literally used tons of DDT on the campus and birds were dying all over the place—robins, yellow warblers, and so on. We lost our screech owls, which never came back. We lost all the yellow warblers in front of the Ag. School building and along that drive. I don't know why they should have been lost but we lost them and it took them years and years to come back.

We picked up all the dead robins we could on this campus . . . and we made a calculation of what the robin mortality was and it was on the order of about 88 to 89 percent. And then our study on the communities of Wauwatosa, Reedville, and Shorewood [Milwaukee suburbs] showed bird populations affected in a variable fashion. What made sense ultimately was that the surviving bird populations had densities that were inversely proportional to the number of pounds of DDT sprayed per acre. The possibility that this relationship occurred by chance was on the order of one in a thousand. So it was absolutely clear that DDT had a profound effect on killing birds the first time it was being used and where it was being used for the third time it left bird populations decimated. Shorewood . . . was essentially a community which had experienced silent spring because the robin population was down an indicated 98 percent.

The effects of the sprays were simply spectacular, they could be seen by householders. The scientists, who don't spend much time at home and don't necessarily live in the sprayed communities, tended to pooh-pooh these things but everything that I saw at that time indicated that the women who complained were not exaggerating. I think some of the things that Miss Carson came out with were exaggerated because these involved things that I simply could not test, such as the effect of these sprays on migratory birds that passed through these communities. That is a very difficult thing to get at, and all I was getting at was the breeding birds. I did get delayed effects, which had already been reported. Barker at the Illinois Natural History Survey found that foliar sprays on elm trees on the Urbana campus resulted in leaves that became contaminated with DDT and dropped to the ground where often they were eaten by earthworms and robins then ate the earthworms and you had robins dying the following spring. Now in Madison on

Capitol Drive on the west side of town trees were sprayed in the fall for Dutch elm disease and the robin mortality took place in the following spring also.

Then two California investigators, Hunt and Bischoff, reported that a curious die-off of grebes at a place called Clear Lake had finally been traced, after years of frustration, not to a disease but to the application of a chemical related to DDT, called DDD, which was placed in the lake for the control of gnats and midges. It was placed in the lake in very low concentrations using extremely conservative applications but to everybody's astonishment the material concentrated identically to what had been found in robins and earthworms at Urbana and the grebes died off.

We just went along like this for several years—Rachel Carson's book out, full of truth, half-truths, and untruths as far the wildlife was concerned, and I have nothing to say about the human health thing—and then in 1962 I heard that the peregrine falcons in the eastern part of the United States, in the northeastern part of the United States, had not raised a single young. I passed that off as a rumor, but . . . the following June a paper comes out in a British journal reporting that the peregrine falcon population in Britain had collapsed and that it possibly could be due to insecticides. I had sent a warning letter to the president of the National Audubon Society saying that if our North American peregrines were undergoing the same phenomenon as in Britain then we indeed were in trouble. I managed to get a grant and the next spring . . . we sent two young historical naturalists named D. D. Berger and Charles Sindelar on a 14,000 mile trip around the eastern part of the United States to look at these birds at nesting sites where they had been known to occupy a cliff for years and years, sometimes dating back to the 1860s, and the thought was that I could direct them to the identical sites that were being used in the 1930s when I did my research on the birds and they could . . . see whether there was any change in the population. They had to cover a region from Alabama all the way to Maine. They started out late in the spring in order to get these eyries, as they are called, when there would be young present. So what they were really censusing was reproductively successful pairs. I was after them . . . to be sure and get to the wilderness sites and not to go to the ones where human occupancy might have destroyed or biased the picture. We could not pull the numbers out of a hat or a table of random numbers but it was a fairly random sample just the same and . . . they had failed to see a single peregrine falcon and they reported that the population was effectively gone. They saw excrement on two sites, which suggested that the birds had been there either early that spring or the preceding year but not a single bird was fledged.

I was spending several months in Europe and meeting the peregrine specialists over there, people in France, Switzerland, Germany and Finland, and the thinking over there was almost as bad. Disaster in all these places. So I came back to get this report from these two fine young men and, aware of what was going on in Europe, I realized these things were fitting right together like a beautiful mesh but we did not understand what had happened to the American birds. They were simply gone and no one seemed to have been studying them. I tried then to link together these people who had been doing some studies of the peregrine falcon on both continents and to run a conference which would review the data and give us leads on what further research needed to be done and that was to take place in 1965. And I wanted to get the government agencies to finance this thing and to keep it out of any possible bias of conservation agencies. Now my good friends in the National Audubon Society, including president Carl Buchheister, were at this point already coming out strongly against pesticides, but I was not in a position to do so.

There is a distinction to be made here between a conservationist and a scientist. A conservationist can afford to take positions that do not necessarily have to be completely documented but as a scientist I could not do this. It was probably perfectly OK for people like Dr. Buchheister and Dr. Peterson, Roger Tory Peterson, to take these firm positions but it was not possible for me to do this, because the trap I was in and a lot of other ornithologists was that we wanted to do research on this thing but we did not want to do research that proved that we were right. That is *not* research. When you are doing research you are testing a hypothesis which holds that the events that you think took place may have occurred only by chance. So I was not by any means ready to come out against DDT at this point. In 1965 we made a conference, with a lot of people who could give us the picture not only on the peregrines but also the North American picture on other birds. What emerged . . . was that indeed the reproductive failures of peregrine falcons on two continents were identical and that the birds were breaking and eating their own eggs. Now the Americans did not realize this when the conference took place but the British convinced them that reports on disappearing eggs had to be this, and we did get some broken eggs in certain eyries. There were more eggshells and fragments left from the British than the American falcons. At any rate we were all convinced the same phenomenon was taking place and there were somewhat similar things taking place, not quite as spectacularly, in other species like ospreys and bald eagles, but why? There was not a single North American peregrine egg run in the laboratory for pesticide analysis at that time, so we were hurting. The British had a much better picture than we did. They had laboratory results.

The North American people knew what to do when the conference was over. Two of them immediately went home and floated a grant application to appropriate agencies, got into the field the next spring, not in the United States, where the birds were so scarce, but in Canada and in Alaska where they were still very numerous, and in some cases shot the birds and in other cases trapped them and took fat samples from them and they found very, very high levels. It took some time to find the eggshell breakage but by the summer of '66, as I was attempting to summarize this conference, I came up with ten ideas as to what might be involved and two of them involved a change in the calcium metabolism of the birds to the effect that the eggshells actually got thinner. I sent these ideas out to two colleagues and then one of them, an Englishman, said he had been quietly measuring the thicknesses of eggshells in the egg collectors' collections in Britain and the eggs indeed have gotten thinner and the change started in 1947. So that following spring . . . I sent a student around the United States to look at museum collections and . . . the few private egg collections we still had. And he found the same thing the Englishman found, in fact he found that in 1947 this change in shell thickness in the peregrine had occurred in California and in Massachusetts. We also got herring gull eggs collected in various parts of the eastern half of the country, Lake Superior, Lake Michigan, from Maine, Block Island . . . and we ran those for residue levels in the eggs and we ran them for eggshell thickness and we found a perfectly inverse correlation—the higher the residue levels in the eggs the thinner the eggshell. And analyzed statistically the indications that this could have occurred by chance were one in a thousand times. And *that* convinced me at that point, which I believe was June of '68, that DDT indeed was the chemical that was causing these things.

15

Backfire in the War against Insects

ROBERT S. STROTHER

There is mounting evidence that massive aerial spraying of pesticides may do more harm than good. Until the full results are known, all concerned should heed the warning: "Proceed with caution!"

The United States is engaged in an intensive war against destructive insects. The weapons employed are powerful and widespread, and so is the controversy they have engendered. Billions of pounds of poisons were broadcast over 100 million acres of cropland and forest. More billions of pounds are being spread across the nation this year—against spruce budworm in northern forests, grasshoppers in nine million acres of wheatland in the Midwest, white fringed beetle in the Southeast; against sand flies, gnats, Japanese beetles, corn borers and gypsy moths. The U.S. Department of Agriculture is only one of the large-scale users of insecticides. State, county and even local groups also employ them, sometimes in cooperation with USDA, sometimes alone.

The new insecticides, often used as massive sprays from planes, kill birds, fish and animals along with insects of all kinds, good as well as bad. The costs of the campaigns in money, destruction of wildlife and possible harm to human health are not adequately known. The need for them is hotly challenged and hotly defended.

Says Dr. George J. Wallace, Michigan State University zoologist, "The current widespread program poses the greatest threat that animal life in North America has ever faced—worse than deforestation, worse than ille-

Reader's Digest 74 (June 1959), 64–69.

gal shooting, worse than drainage, drought, oil pollution; possibly worse than all these decimating factors combined."

The USDA points to the eradication of the Mediterranean fruit fly in Florida as an example of what spray programs can accomplish. It also cites thousands of letters from grateful farmers all over the country. "Benefits to agriculture and the public," says USDA, "far outweigh damages that have occurred. Claims of wildlife destruction are greatly overstated."

Nobody knows for certain just how much damage is done, but there have been indications. In Florida, when a 2000-acre tidal marsh was treated with dieldrin for sandfly larvae, a biologist was on hand. His report: "The fish kill was substantially complete. The immediate over-all kill was twenty to thirty tons of fish, or about 1,175,000 fish, of at least thirty species. The larger game and food fish succumbed most rapidly. Then crabs devoured them; next day the crabs themselves were dead."

Large numbers of dead and dying birds—mostly robins—were found on the campus of Michigan State University in the spring of 1955. Indications were that death was due to insecticide poisoning, and subsequent investigation disclosed the chain of events. In the previous year, campus elms had been sprayed against bark beetles, carriers of the fungus which causes Dutch elm disease. Earthworms eating the leaf litter accumulated DDT in their viscera. When spring came and the worms emerged from the thawed ground, the robins ate the worms and died.

The annual elm spraying was continued. Its effect on reproduction among surviving robins was startling. In the spring of 1957 zoologists searched the 185-acre North Campus for nests. Only six were found. Of these, five produced *no* young and the fate of the sixth was undetermined. Late in June a three-day search for young robins found only one—all this in an area that in 1954 had produced, on the basis of sampling counts, at least 370 young robins.

The first public outcry against massive spraying arose in 1957 during the USDA campaign against the gypsy moth in southern New York. Planes flew over at low levels, discharging a fog of DDT-impregnated kerosene on three million acres, including densely populated communities in Westchester County and on Long Island. Commuters awaiting their trains were sprayed, as were dairy farms, ponds, vegetable gardens and children, some of them three times.

Tempers flared, and fourteen citizens charging careless use and official arrogance went into federal court demanding an injunction against aerial spraying. After an extended hearing their application was denied.

Another and louder outcry was in the making. According to reports, twenty-seven million acres in nine Southern states from eastern Texas to South Carolina were "teeming" with South American fire ants. These quarter-inch-long ants, it was said, had captured much of the South's best farm land and were eating their way north and west, sucking plant juices, killing young wildlife and swarming in vicious assault on men in the fields. Their onslaught, if unchecked, might not stop short of California and Canada.

To combat the menace Congress voted an emergency appropriation of $2,400,000 for USDA. Plant Pest Control Crews, without prior field testing, started an aerial broadcast of heptachlor, a powerful chlorinated hydrocarbon of the DDT family. They treated 700,000 acres before the appropriation ran out. The USDA now has asked Congress for another $2,400,000 for the second step in a long-range poisoning program. Does the fire-ant threat justify this costly campaign?

To get a firsthand view I went to Alabama, where fire ants have flourished for forty years, and talked with people for, against and in the middle of the program. Some remarkable facts emerged.

The foremost is this: the fire ant is not a serious crop pest; it may not be a crop pest at all. Dr. F. S. Arant, chief of the zoology-entomology department at Alabama Polytechnic Institute, told me, "Damage to crops by the imported fire ant in Alabama is practically nil. This department has not received a single report of such damage in the past five years. No damage to livestock has been observed. The ant eats other insects, including the cotton boll weevil. It is a major nuisance, but no more."

Though USDA circular No. 350 asserted that the imported fire ants often attack newborn calves and pigs, are fond of quail and chase brooding hens off their nests to eat their chicks,

WHAT ONE BIRD CAN DO

A house wren feeds 500 spiders and caterpillars to its young during one summer afternoon.

A swallow devours 1000 leafhoppers in twelve hours.

A pair of flickers consider 5000 ants a mere snack.

A Baltimore oriole consumes seventeen hairy caterpillars a minute.

A brown thrasher can eat 6180 insects in one day.

—*Garden Club of America Conservation Committee*

researchers at Alabama Polytechnic could rarely induce fire ants, even starving ones, either to eat plants or attack young animals. Instead, the insects became cannibalistic and ate one another.

Farmers and cattlemen detest the fire ants because their ugly two-foot-high mounds clog mowing machines, and the ants bite when the farmer gets down to clear the blades. But none of the farmers I talked with had suffered any crop damage from fire ants. To control them, all farmers have to do is pour insecticides into the individual mounds or disk insecticides into the soil along with fertilizers—two successful, low-cost methods developed by the agricultural experiment stations of Alabama and Mississippi.

Last September, fifty-two experts, including a USDA contingent of five, were present at a fire-ant research meeting in Auburn, Ala. Dr. L. D. Newson of Louisiana State University challenged anyone there to go on record as saying that the fire-ant eradication program could be justified by damages to crops or animals. There was no answer.

What was the effect of the 1957–58 fire-ant campaign on wildlife? Dan Lay, Texas wildlife biologist, reported from Hardin County: "On May 12, before the poisoning, the fields were noisy with birds singing for territorial establishment. Dickcissels, red-winged blackbirds and meadow larks were building nests and laying eggs. Forty-one nests with eggs were found in one clover field."

Then the planes came, scattering tiny pellets of clay containing 10 percent heptachlor. The poison covered the ground, seven to twelve sugar-size granules to the square inch. The birds ate poisoned insects, pulled worms through poisoned soil or absorbed the poison through their feet. Within a day they began to tremble, went into convulsions and died. Orphan broods hatched and died in their nests. By June 3 only three of the forty-one nests in the clover field remained occupied. Birds along the roads were reduced 95 percent.

It was the same in other areas sprayed: quail and killdeer wiped out; doves, woodpeckers, snipe, mockingbirds, cardinals, woodcock, hawks, wild turkeys, shrikes and many other species almost exterminated.

Animals died, too. A raccoon which had been seen rolling frenziedly in the road was later found dead by the roadside. Four fox pups were found dead in their den, poisoned by food brought in by their mother. Fish, turtles, snakes, rabbits, opossums, squirrels, armadillos were killed.

Today fish and game commissions in most of the afflicted states, finding the cure worse than the disease, have demanded a halt to aerial spraying.

"It's like scalping yourself to cure dandruff," said Clarence Cottam, former official in the Fish and Wildlife Service.

"Sickening," said Charles Kelley of the Alabama Conservation Department. "These people can kill more game in a month than our department can build up in twenty years." Kelley handed me one of the USDA warnings given people whose lands are about to be doused:

> Cover gardens and wash vegetables before eating them; cover small fish-ponds; take fish out of pools and wash pools before replacing the fish; don't put laundry out; keep milk cows off treated pastures for 30 days, and beef cattle 15 days; cover beehives or move them away; keep children off ground for a few days; don't let pets or poultry drink from puddles.

"How can any official read that and still say the losses of wildlife are insignificant?" he demanded.

Last year, under the prodding of Rep. Lee Metcalf of Montana, whose interest grew out of the wholesale destruction of fish in the Yellowstone River following a mishap in spraying DDT, Congress conducted hearings on the pesticide problem and set aside $125,000 for studies to learn what we are doing to fish and wildlife. All witnesses agreed basic research was badly needed. They raised many unanswered questions.

Do repeated small doses of the poisons impair the reproductive ability and lower the survival rate of the young in animals and birds? Most of the new chemicals retain their killing power in the soil for three years at least. Can they still kill after five years? Ten? Nobody knows.

What of the microorganisms that create soil fertility in the first place? What of the bottom organisms in streams and bays, on which marine life feeds? Do they accumulate the poisons? Since pesticides kill mice-eating hawks, owls and foxes as well as rodents, and beneficial as well as harmful insects, may we not find ourselves without natural allies in the war on pests, and become wholly dependent on ever stronger chemicals?

What about insects' developing immunity, just as some germs have become immune to penicillin? The housefly and the mosquito were the initial targets for the new sprays. Now some common species of these insects are 1800 times more resistant to DDT than were their ancestors of a few years ago. Are we trading a costly temporary victory over other pests for disaster in the form of super-insects later on? Nobody knows.

And finally there is the greatest question of all: how serious are the hazards to human health? Doctors are increasingly troubled by the possibility that DDT and its much more poisonous descendants may be responsible for the rise in leukemia, hepatitis, Hodgkin's disease and other degenerative diseases.

It may take years to find the answers to some of these questions. But one thing seems plain enough: aerial spraying needlessly kills wildlife and should not be done except in small areas and real emergencies.

Our forests flourished without chemical help through eons of time, and man has practiced agriculture with reasonable success for 100 centuries of recorded history. The new pesticides have been in general use for fifteen years. "Surely," says Dr. Fairfield Osborn, noted conservationist, "we would be wise to halt massive spraying until we know what effects the toxins are having on ourselves and our animal co-heirs to this planet."

Dr. Wallace has made a grim prophecy: "If this and other pest-eradication programs are carried out as now projected, we shall have been witnesses, within a single decade, to a greater extermination of animal life than in all the previous years of man's history on earth."

This may be a wildly pessimistic view. Nobody knows. But why risk it?

PART 4

THE STORM OVER *SILENT SPRING*

Rachel Carson changed the emerging debate over DDT, partly by putting together scattered research results into a coherent case but mainly by looking beyond human health to environmental safety and arguing not for new laws but a new relation between humans and nature. The scientific knowledge and literary skill that made *Silent Spring* a best-seller, the center of controversy, and then an environmental classic came from her whole career. Interested in nature from childhood, she started as an English major in college, then switched to biology, received a master's degree, and worked for the Fish and Wildlife Service from 1936 to 1951 as researcher and editor, eventually becoming the agency's main scientific editor and writer of many of its popular publications. When the success of her second book, *The Sea Around Us*, allowed her to become a full-time nature writer, she continued to follow the DDT story.

The first selection here, though, comes not from her book but a newspaper. It is an account of the thalidomide disaster, which became public in the summer of 1962, as the *New Yorker* printed long sections of her forthcoming manuscript. Doctors traced an epidemic of deformed babies to a sleeping pill and anti-nausea medicine used in Europe and under review in the United States that, taken during a critical period of pregnancy, caused characteristic birth deformities. Pictures, including one of a baby girl in a dress, using her feet to play with a wristwatch because she had no arms, horrified the public. This article, describing the pressure on Dr. Kelsey to approve the drug for use in the United States, points up, indirectly, a key issue in the pesticide controversies. Who decided, and on what grounds, what was "safe" for people to take into their bodies?

The chapter from *Silent Spring* that follows, "A Fable for Tomorrow," spoke of safety and damage in a larger frame than immediate human health, wildlife, and the land around us. It did not, as Carson made clear, describe any single town, but it dramatized problems that had occurred across the country and suggested our future if we continued what she characterized as our reckless use of pesticides. It set the agenda for the rest of her book, which showed the scientific evidence behind her disturbing vision and presented another approach to insect control, based on biological and ecological knowledge, on living with nature rather than conquering it.

The uproar that followed the book's publication caused President Kennedy to ask the President's Science Advisory Committee for a report. The members had no particular expertise in this issue but were respected scientists from several disciplines, picked by the president to advise him on matter of national importance. Their assessment, *Use of Pesticides*, weighed known and possible dangers to human health and environment against the expense of regulation and the price of food. The stand they took now seems cautious, but they had a wider perspective than earlier reports, for they looked, as earlier groups had not, beyond the people who sprayed chemicals to all humans, wildlife populations, and even ecosystems and considered dangers besides obvious illness. Their bureaucratic prose also revealed a skepticism about the unmixed blessings of technology not visible a decade before.

The next two pieces argued the case for pesticides. Robert H. White-Stevens, who worked for a chemical manufacturer, rallied pesticide manufacturers and users in "Communications Create Understanding" with an appeal to reason and public discussion, calling on them to make the public aware of the benefits of pesticides. Edwin Diamond, a popular writer who had tried collaborating with Carson until it became clear how far apart their views were, argued, in "The Myth of the 'Pesticide Menace,'" that she exaggerated and distorted some facts, overlooked others, but (most important) ignored our need for pesticides. Both men saw human life as a struggle against nature in which technology provided our greatest weapons and believed humans had only two choices: to conquer, which meant following current practices, or to surrender to the forces of nature, which would mean catastrophe.

The final piece goes over the legal maneuvering that went into the national ban on DDT. It summarized the issues, scientific and political, that emerged from the hearings and decisions but also pointed out that ending DDT use and then DDT manufacture required more than a single, decisive action. The ban required judgments on safety but also adjustments of policy to political interests.

16

"Heroine" of FDA Keeps Bad Drug Off Market

MORTON MINTZ

This is the story of how the skepticism and stubbornness of a government physician prevented what could have been an appalling American tragedy, the birth of hundreds or indeed thousands of armless and legless children.

The story of Dr. Frances Oldham Kelsey, a Food and Drug Administration medical officer, is not one of inspired prophecies or of dramatic research breakthroughs.

She saw her duty in sternly simple terms, and she carried it out, living the while with insinuations that she was a bureaucratic nitpicker, unreasonable—even, she said, stupid. That such attributes could have been ascribed to her is, by her own acknowledgement, not surprising, considering all of the circumstances.

What she did do was refuse to be hurried into approving an application for marketing a new drug. She regarded its safety as unproved, despite considerable data arguing that it was ultra safe.

It was not until last April 19, months after the application was filed with the FDA, that the terrible effects of the drug abroad were widely reported in this country. What remains to be told is how and why Dr. Kelsey blocked the introduction of the drug before those effects were suspected by anyone.

Washington Post 15 July 1962, 1.

Dr. Kelsey invoked her high standards and her belief that the drug was "peculiar" against these facts:

The drug had come into widespread use in other countries. In West Germany, where it was used primarily as a sedative, huge quantities of it were sold over the counter before it was put on a prescription basis. It gave a prompt, deep, natural sleep that was not followed by a hangover. It was cheap. It failed to kill even the would-be suicides who swallowed massive doses.

And there were the reports on experiments with animals. Only a few weeks ago, the American licensee told of giving the drug to rats at doses six to sixty times greater than the comparable human dosage. Of 1510 offspring none was delivered with "evidence of malformation."

In a separate study one rat did deliver a malformed offspring, but the dosage had been 1200 times the usual one. Rabbits that were injected with six times the comparable human dose also were reported to have produced no malformed births.

Recently, the FDA publicly decried the "excessive contacts" made with its personnel by pharmaceutical manufacturers who are anxious to speed the agency's handling of new drug applications.

MANY REQUESTS

So it was not at all surprising that dozens of contacts were made with Dr. Kelsey by representatives of the American licensee for thalidomide, the chemical name for the sedative. They had what they strongly believed was a clear and overwhelming case—but Dr. Kelsey delayed, and delayed, and delayed.

They visited her in her drably furnished, bare-floor office in an eyesore Tempo [temporary building] on Jefferson Drive, S.W. They phoned. They submitted a flow of reports and studies. It was apparent that substantial investments and substantial profits were at stake. And all of this was routine.

The application had come to Dr. Kelsey—simply because it was her turn to take the next one—in September, 1960.

The European data left her "very unimpressed." In an interview, she said that she had "lived through cycles before" in which a drug was acclaimed for a year or two—until harmful side effects became known.

And, she said, she could not help regarding thalidomide as "a peculiar drug." It troubled her that its effects on experimental animals were not the same as on humans—it did not make them sleepy.

SAME QUESTIONS

Could there be danger in those few people whose systems might absorb it? Could there be a harmful effect on an unborn child whose mother took it? (In other countries obstetricians were innocently prescribing it as an antiemetic for pregnant women.)

Dr. Kelsey regarded the manufacturer's evidence of thalidomide's safety as "incomplete in many respects." The drug was not, after all, intended for grave diseases, or for the relief of intolerable suffering, but primarily for sleeplessness, for which many drugs of known safety were already on the market.

All of this being so, she saw no need either to hurry or to be satisfied with the approach that, nine chances out of ten, it's safe. She was determined to be certain that thalidomide was safe ten times out of ten, and she was prepared to wait forever for proof that it was.

When the sixty-day deadline for action on the application came around, Dr. Kelsey wrote the manufacturer that the proof of safety was inadequate. Perhaps with an understandable sense of frustration the manufacturer produced new research data, new reasons for action. Each time a new sixty-day deadline drew near, out went another letter: insufficient proof of safety.

UPHELD BY SUPERIORS

Dr. Kelsey's tenacity—or unreasonableness, depending upon one's viewpoint—was upheld by superiors, all the way.

Although she takes her work seriously indeed, her contacts with applicants are, in her words, "usually amiable. We see their point, and they see ours. But the responsibility for releasing a drug is ours, not theirs." And that is the responsibility she would not forget.

In February, 1961, she chanced to read, in a British medical journal, a letter from a British doctor questioning whether certain instances of peripheral neuritis—a tingling and numbness in the feet and the fingers that is sometimes irreversible—might not be due to intake of thalidomide. To her this was a danger signal.

She called the letter to the attention of the applicant. His investigators reported that the incidence was apparently negligible, one case among

300,000 adult users. Six months later, Dr. Kelsey said, the incidence among adults who took thalidomide regularly for months at a time was found to be one in 250.

But neither she nor the applicant yet had the slightest inkling that the drug could be responsible for the birth of malformed babies. That awful circumstantial evidence became known to the applicant—in a cablegram from Europe—on Nov. 29, 1961.

APPLICATION WITHDRAWN

He reported it to Dr. Kelsey early the next day. Although this was followed by a formal withdrawal of the application, as late as last month the applicant described the birth abnormalities an "alleged effects" of thalidomide.

The story begins in 1954, six years before Dr. Kelsey, a pharmacologist as well as a physician, went to work in the FDA's Bureau of Medicine. She and her husband, F. Ellis Kelsey, a pharmacologist who is now a special assistant to the Surgeon General of the Public Health Service, came here from the faculty of the University of South Dakota School of Medicine.

For the account that follows the primary sources were Dr. Kelsey and reports by Dr. Helen B. Taussig to a medical meeting in April and in the June 30 issue of the *Journal of the American Medical Association.*

Dr. Taussig, professor of pediatrics at the Johns Hopkins School of Medicine in Baltimore, went to West Germany in January to investigate the relationship between thalidomide and an enormous increase in the birthrate of malformed infants.

Eight years ago a West German manufacturer conceived of the drug, synthesized it—and discarded it after discerning no effect in test animals. In 1958 another West German firm also developed thalidomide and found it to be, by all indications, the best sleeping compound ever devised.

LARGE SALE

The sale was tremendous. It even came to be used for grip, neuralgia, asthma, in cough medicines and to calm children before they were given electroencephalograms.

In Germany it was marketed as Contergan, in the British Commonwealth as Distaval, in Portugal as Softenon. Dr. Kelsey's native Canada accepted it on April 1, 1961, for manufacture by one firm under the name Talimol and by another firm, the William S. Merrell Co. of Cincinnati, under the name Kevadon. It was the 134-year-old Merrell firm that was seeking to market Kevadon as a prescription drug in the United States.

At that time—April, 1961—West German investigators were desperately groping for an explanation of an unprecedented outbreak of phocomelia, a malformation hitherto so rare that it isn't even listed in some medical dictionaries. An eighty-six year old Gottingen specialist in human deformities told Dr. Taussig that he had in his whole lifetime "seen as many individuals with two heads as he had with phocomelia."

Usually, phocomelia deprives its victims of one arm. Rudimentary fingers that look, said Dr. Taussig, "like the flippers of a seal" arise from the stub below the shoulder.

CLINIC CASES

In eight West German pediatric clinics there were no cases at all between 1954 and 1959. In 1959 there were twelve, in 1960 there were eighty-three, in 1961 there were 302.

These were not the ordinary textbook cases. Not just one arm was affected. These children were without both arms or without both legs, or without three limbs, or they were without any limbs at all.

In some the external ear was missing and hearing was grossly impaired. There were deformities of the eyes, esophagus and intestinal tract; and even this is not a complete list.

Once the suspected link with Contergan was established, Contergan was taken off the West German market. The expectation is that the last mothers who could have taken it during early pregnancy, the danger period, will be delivered in August. The estimates are that by the end of next month the total of deformed children born in West Germany will be between 3500 and 6000. Two out of three are expected to live. Most are apparently of normal mentality.

The drug was withdrawn from the British market five days after the withdrawal in West Germany. *The Guardian*, Manchester, has predicted that August will see the birth of 800 deformed English children. The Ministry of Health has begun to fit fifty victims with artificial limbs.

EIGHT IN CANADA

An article prepared for the May 9 issue of *Maclean's Magazine* said that at the time of writing eight victims of phocomelia had been born in Canada, two of them to physicians' wives who had used "samples of thalidomide donated to their husbands."

Because the Department of Health did not order thalidomide withdrawn from sale until March 2, *Maclean's* said the last Canadian casualties are not expected until November.

The cause of the West German outbreak was hard to trace. Hereditary factors, blood incompatibility between parents, abnormal chromosomes, radioactive fallout, X-rays, detergents, food preservatives—all of these things, and more, were suspected, checked, and discarded as possibilities.

A Hamburg pediatrician, Dr. Widukind Lenz, made preliminary studies showing that about 20 percent of the mothers who brought deformed infants to his clinic had taken Contergan. Dr. Taussig wrote: "On Nov. 3, 1961, it occurred to him that Contergan was the cause. He requestioned his patients and the incidence promptly rose to about 50 percent. Many of the patients said they had considered the drug too innocent to mention it on the questionnaire."

MAKER WARNED

"On Nov. 15 he warned Grunenthal (the manufacturer) that he suspected Contergan was the cause and that the drug should be withdrawn."

Five days later, at a pediatric meeting in Dusseldorf, he reported his suspicions and his actions but did not name the drug. That night, Dr. Taussig related, " a physician came up to him and said, 'Will you tell me confidentially, is it the drug Contergan? I ask because we have such a child, and my wife took Contergan.'

"A couple of days later it was generally known that Contergan was the drug under suspicion. On Nov. 26 Grunenthal withdrew the drug from the market. On Nov. 28 the Ministry of Health issued a firm but cautious and widely publicized statement that Contergan was suspected to be a major factor in the production of phocomelia."

Dr. Taussig reported that an Australian physician, Dr. W. G. McBride, saw three severe cases in April, 1961, and three more in October and November. "He found that all six mothers had taken Distaval in early pregnancy," the *Journal* article said.

In Stirlingshire, Scotland, Dr. A. L. Spiers saw 10 severe phocomelia victims during 1961 and ultimately "obtained positive proof that eight out of ten of these patients had taken Distaval."

DIFFICULT CONNECTION

Making the connection—which some physicians say is not conclusively established—was extraordinarily difficult.

Dr. Lenz, for example, had to contend with the lack of records during the time when Contergan was sold without prescription, and with his patients' natural difficulty in recalling if and precisely when they had taken a sleeping pill months earlier.

"In one instance," Dr. Taussig wrote, a doctor "swore the mother had not received Contergan. He had prescribed an entirely different sedative. On investigation at the pharmacy . . . Dr. Lenz found the prescription was stamped 'drug not in stock, Contergan given instead.'"

Dr. Taussig said the investigations of Dr. Lenz in particular indicate that the embryo is endangered if a mother takes thalidomide within about twenty to forty days after conception, a time when she may not even know that she is pregnant.

He believes that during that sensitive period the chances that a mother who has taken the drug will deliver a deformed baby are at least two in five.

COMPANY VIEW

The Merrell firm says that conclusive proof is lacking for such assumptions and cites a clinic in Kiel at which, Merrell reported, half of the deformed children were delivered to mothers who probably had not taken thalidomide.

"Everyone admits," Dr. Taussig wrote, "that no information is available concerning how many women may have taken the drug in the sensitive period and have had a normal child."

Dr. Kelsey said the molecular complex of thalidomide is being broken down and studied in an effort to determine the causative agent in thalidomide.

In all of this Dr. Taussig sees compelling reason for caution in the use of new drugs by women of child-bearing age. A Canadian physician interviewed by *Maclean's* said, "There is too much demand on the part of the public for relief of mild or even moderately severe symptoms. People won't put up with even the slightest discomfort or headache; they demand medication from their doctor. If they can't get it from one, they'll go to another."

Dr. Taussig also wants the 1938 Food and Drug Act strengthened to provide greater assurance that new drugs will not harm unborn children. But to Assistant FDA Commissioner Winton B. Rankin, the significant thing about the law is that it gave Dr. Kelsey the weapons she needed to block the marketing of thalidomide in the United States.

"The American public," he said, "owes her a vote of thanks.'"

The 47-year-old Dr. Kelsey lives at 4811 Brookside Drive, Chevy Chase, with her husband and daughters, Susan, fifteen, and Christine, twelve. She is grateful for the praise—but recognizes that, had thalidomide proved to be as safe as the applicant believed, "I would have been considered unreasonable."

She intends to go on "playing for that tenth chance in ten" to assure safety in new drugs "to the best of my ability." For twenty years she taught pharmacology. She knows the dangers, and she has not the slightest intention of forgetting them.

17

A Fable for Tomorrow

RACHEL CARSON

There was once a town in the heart of America where all life seemed to live in harmony with its surroundings. The town lay in the midst of a checkerboard of prosperous farms, with fields of grain and hillsides of orchards where, in spring, white clouds of bloom drifted above the green fields. In autumn, oak and maple and birch set up a blaze of color that flamed and flickered across a backdrop of pines. Along the roads, laurel, viburnum and alder, great ferns and wildflowers delighted the traveler's eye through much of the year. Even in winter the roadsides were places of beauty, where countless birds came to feed on the berries and on the seed heads of the dried weeds rising above the snow. The countryside was, in fact, famous for the abundance and variety of its bird life, and when the flood of migrants was pouring through in spring and fall people traveled from great distances to observe them.

Then a strange blight crept over the area and everything began to change. Some evil spell had settled on the community: mysterious maladies swept the flocks of chickens; the cattle and sheep sickened and died. The farmers spoke of much illness among their families. In the town the doctors had become more and more puzzled by new kinds of sickness appearing among their patients. There had been several sudden and unexplained

Silent Spring (Boston: Houghton Mifflin, 1962), 1–3.

deaths, not only among adults but even among children, who would be stricken suddenly while at play and die within a few hours.

There was a strange stillness. The birds, for example—where had they gone? Many people spoke of them, puzzled and disturbed. The feeding stations in the backyards were deserted. The few birds seen anywhere were moribund; they trembled violently and could not fly. It was a spring without voices. On the mornings that had once throbbed with the dawn chorus of robins, catbirds, doves, jays, wrens, and scores of other bird voices there was now no sound; only silence lay over the fields and woods and marsh.

On the farms the hens brooded, but no chicks hatched. The farmers complained that they were unable to raise any pigs—the litters were small and the young survived only a few days. The apple trees were coming into bloom but no bees droned among the blossoms, so there was no pollination and there would be no fruit.

The roadsides, once so attractive, were now lined with browned and withered vegetation as though swept by fire. These, too, were silent, deserted by all living things. Even the streams were now lifeless. Anglers no longer visited them, for all the fish had died.

In the gutters under the eaves and between the shingles of the roofs, a white granular powder still showed a few patches; some weeks before it had fallen like snow upon the roofs and the lawns, the fields and streams.

No witchcraft, no enemy action had silenced the rebirth of new life in this stricken world. The people had done it themselves.

This town does not actually exist, but it might easily have a thousand counterparts in America or elsewhere in the world. I know of no community that has experienced all the misfortunes I describe. Yet every one of these disasters has actually happened somewhere, and many real communities have already suffered a substantial number of them. A grim specter has crept upon us almost unnoticed, and this imagined tragedy may easily become a stark reality we all shall know.

What has already silenced the voices of spring in countless towns in America? This book is an attempt to explain.

18

Use of Pesticides

PRESIDENT'S SCIENCE ADVISORY COMMITTEE

Evidence of increasing environmental contamination by pesticide chemicals has generated concern which is no longer limited to citizens of affected areas or members of special-interest groups. During two decades of intensive technical and industrial advancement we have dispersed a huge volume of synthetic compounds, both intentionally and inadvertently.

Today, pesticides are detectable in many food items, in some clothing, in man and animals, and in various parts of our natural surroundings. Carried from one locality to another by air currents, water runoff, or living organisms (either directly or indirectly through extended food chains), pesticides have traveled great distances and some of them have persisted for long periods of time. Although they remain in small quantities, their variety, toxicity, and persistence are affecting biological systems in nature and may eventually affect human health. The benefits of these substances are apparent. We are now beginning to evaluate some of their less obvious effects and potential risks.

Although acute human poisoning is a measurable and, in some cases, a significant hazard, it is relatively easy to identify and control by compari-

Washington: Government Printing Office, 1963.

son with potential, low-level chronic toxicity which has been observed in experimental animals.

The Panel is convinced that we must understand more completely the properties of these chemicals and determine their long-term impact on biological systems, including man. The Panel's recommendations are directed toward these needs, and toward more judicious use of pesticides or alternate methods of pest control, in an effort to minimize risks and maximize gains. They are offered with the full recognition that pesticides constitute only one facet of the general problem of environmental pollution, but with the conviction that the hazards resulting from their use dictate rapid strengthening of interim measures until such time as we have realized a comprehensive program for controlling environmental pollution.

The worldwide use of pesticides has substantially increased since the development of DDT and other chlorinated hydrocarbons in the early 1940's. It is estimated that 350 million pounds of insecticides alone were used in the United States during 1962. They are distributed annually over nearly ninety million acres (about one acre out of twenty within the forty-eight contiguous states). Thus the land area treated with pesticides is approximately one acre of twelve within the forty-eight states. About forty-five million pounds are used each year in urban areas and around homes, much of this by individual homeowners. The annual sale of aerosol "bug bombs" amounts to more than one per household. Other compounds, such as fungicides, also are used in substantial tonnage.

In recent years we have recognized the wide distribution and persistence of DDT. It has been detected at great distances from the place of application and its concentration in certain living organisms has been observed. DDT has been found in oil of fish that live far at sea and in fish caught off the coasts of eastern and western North America, South America, Europe, and Asia. Observed concentrations have varied from less than 1 part per million (ppm) to more than 300 ppm in oil.

Residues of DDT and certain other chlorinated hydrocarbons have been detected in most of our major rivers, in ground water, in fish from our fresh waters, in migratory birds, in wild mammals, and in shellfish. Small amounts of DDT have been detected in food from many parts of the world, including processed dairy products from the United States, Europe, and South America. The amounts are rarely above Food and Drug Administration (FDA) tolerance limits, but these have probably contributed to the buildup of DDT we now observe in the fat of the people of the United States, Canada, Germany, and England. In the United States, DDT and its metabolites have been found in the fat of persons without occupational exposure at an average of twelve ppm (approximately 100 to 200 mg. of DDT per adult) for the past ten years. In England and Germany, recent studies revealed an average concentration of two ppm in human fat. Data about children are not available.

An important characteristic of several commonly used pesticides is their persistence in the environment in toxic form. The chemical half life of stable chlorinated hydrocarbons in soils, and the time they remain active against some soil insects, are measured in years.

There have been few systematic studies of people occupationally exposed to pesticides. Limited groups of adults occupationally exposed to the more toxic pesticides are also being studied, and there is evidence of neurologic impairment, usually reversible, in those individuals heavily exposed to certain chlorinated hydrocarbons and organic phosphates. Unfortunately, possible long-term effects of other compounds cannot be predicted on the basis of experience with DDT, or even predicted for DDT itself, on the basis of the limited clinical studies available.

The Panel's recommendations are directed to an assessment of the levels of pesticides in man and his environment; to measures which will augment the safety of present practices; to needed research and the development of safer and more specific methods of pest control; to suggested amendments or public laws governing the use of pesticides; and to public education.

In order to determine current pesticide levels and their trends in man and his environment, it is recommended that the Department of Health, Education, and Welfare:

Develop a comprehensive data-gathering program so that the levels of pesticides can be determined in occupational workers, in individuals known to have been repeatedly exposed, and in a sample of the general population. These studies should use samples sufficiently large and properly drawn to obtain a clear understanding of the manner in which these chemicals are absorbed and distributed in the human body.

Cooperate with other departments to develop a continuing network to monitor residue levels in air, water, soil, man, wildlife, and fish.

In order to augment the safety of present practices, it is recommended that:

The Food and Drug Administration proceed as rapidly as possible with its current review of residue tolerances, and the experimental studies on which they are based. Of the commonly used chemicals attention should be directed first to heptachlor, methoxychlor, dieldrin, aldrin, chlordane, lindane, and parathion because their tolerances were originally based upon data which are in particular need of review.

The existing Federal advisory and coordinating mechanisms be critically assessed and revised as necessary to provide clear assignments of responsibility for control of pesticide use. The Panel feels that the present mechanisms are inadequate.

The accretion of residues in the environment be controlled by orderly reduction in the use of persistent pesticides.

As a first step, the various agencies of the Federal Government might restrict wide-scale use of persistent insecticides except for necessary control of disease vectors. The Federal agencies should exert their leadership to induce the States to take similar actions.

Elimination of the use of persistent toxic pesticides should be the goal.

Although data are available on acute toxic effects in man, chronic effects are more readily demonstrated in animals because their generation time is shorter, and thus the natural history of pesticide effects is telescoped chronologically. However, there will continue to be uncertainties in the extrapolation from experimental animals to man, and in the prediction of the nature and frequency of effects in humans on the basis of those observed in other forms of life.

The Panel recommends that toxicity studies include determination of—

Effects on reproduction through at least two generations in at least two species of warmblooded animals. Observations should include effects on fertility, size and weight of litter, fetal mortality, teratogenicity, growth and development of sucklings and weanlings.

Chronic effects on organs of both immature and adult animals, with particular emphasis on tumorigenicity and other effects common to the class of compounds of which the test substance is a member.

Possible synergism and potentiation of effects of commonly used pesticides with such commonly used drugs as sedatives, tranquilizers, analgesics, antihypertensive agents, and steroid hormones, which are administered over prolonged periods.

The Panel recommends expanded research and evaluation by the Department of the Interior of the toxic effects of pesticides on wild vertebrates and invertebrates.

The study of wildlife presents a unique opportunity to discover the effects on the food chain of which each animal is a part, and to determine possible pathways through which accumulated and, in some cases, magnified pesticide residues can find their way directly or indirectly to wildlife and to man.

To enhance public awareness of pesticide benefits and hazards, it is recommended that the appropriate Federal departments and agencies initiate programs of public education describing the use and the toxic nature of pesticides. Public literature and the experiences of Panel members indicate that, until the publication of *Silent Spring* by Rachel Carson, people were generally unaware of the toxicity of pesticides. The Government should present this information to the public in a way that will make it aware of the dangers while recognizing the value of pesticides.

19

Communications Create Understanding

ROBERT H. WHITE-STEVENS

In a democracy such as ours where legislation often is initiated and swayed by public opinion, assiduous attention to the truth and the facts, both in citation and interpretation, is imperative

The single feature of our civilization to which can be attributed the largest credit for the incredible advances of man in the first half of the twentieth century is communication. Communication of information, both visual and verbal, has become virtually instantaneous all over the world; communication of people as travelers has been reduced to hours all over the world; communication of goods and services has become completely worldwide, limited in time or place only by economics and politics. No longer is there separation between or within peoples, except that forcefully imposed by man himself.

The tremendous advantages of such vast communication necessarily invokes concomitant disadvantages. There can no longer be secrets within or among nations or groups of peoples, and although efforts may be made to restrict and classify information, it is at best only a very temporary and transitory restraint. More important, however, is the quick awareness gathered by all peoples of any economic social, scientific or political advantage gained by any one people and their prompt demand for equivalent benefits. This is, perhaps, the most impelling force behind the drive for political, economic, and social emergence of virtually every backward nation on the planet. As a result, the starvation and want that pervades so many of the

Agricultural Chemicals 17 (October 1962), 34 ff.

poorer and less developed nations is no longer regarded as inevitable, but merely intolerable. We have spread the word so effectively to the far corners of the earth of how munificent and luxurious is life in the Western world that every nation on earth is envious and feels entitled to demand its share, and many effectively do. We cannot expect to tell the hungry peoples of the earth of the astounding accomplishments of our agriculture, and announce the unwanted and economically embarrassing surpluses of foodstuffs, without engendering envy and avarice. This is one price of communication.

Another, and perhaps most significant, price we must pay for our modern system of communication lies in the tremendous responsibility it places upon us to use it wisely, truthfully, and with balance and restraint. It is perhaps here where the physical technology of our communication facilities has far outstripped our ethical and intellectual use of such facilities. As Mark Twain said many years ago, "Freedom of speech does not mean one may cry 'Fire!' in a crowded theater," and similarly, in our day, the opportunity to address fifty million U.S. citizens on TV or radio enlarges enormously the responsibility to avoid exaggeration, hyperbole, and misrepresentation. The incredible impact that misplaced emphasis may exert upon the political-economic behavior of a people, and the virtually irrevocable effect such lack of objectivity may have, demand that the responsible speaker or writer meter his words to the facts with careful precision. In a democracy such as ours, where legislation is often initiated and swayed and always executed in accordance with public opinion and acceptance, such assiduous attention to the truth and the facts, both in citation and interpretation, becomes doubly imperative.

Yet, we have seen many cases in recent years where the welfare and livelihood of millions of people have been adversely affected by the deliberate misrepresentation of facts, and where our vast and penetrating communications system has been employed not to inform but to mislead. In former years, when our communications were less rapid and less ubiquitous, correction could the more readily catch up with error and maintain a favorable balance of understanding. Today, however, we often find that a false statement takes off like an ICBM and explodes in TV, radio, newspapers, magazines, and even in books, before a measured estimate of the facts in the matter can be assessed and presented objectively.

At the present time, our industry, among others, is sustaining just such an unjustifiable series of attacks. Small scraps of facts have been misrepresented and misquoted to the general public—usually the urban public that now constitutes the majority of our people—for the deliberate purpose of creating a false alarm and of influencing legislation. In the past, many cases of these specious and mendacious arguments have come from critics with neither knowledge nor responsibility in the fields of agricultural chemistry or food production, and they, therefore, have been dismissed generally as irrelevant nonsense. But, today, some of the statements casting reflection upon the use of agricultural chemicals have emerged from those whose education, training, and recognition by the public are too extensive to be ignored. The fact that their education, experience, and acceptance by the public as authorities places them in a position of responsibility to the public seems, for the moment at least, to have escaped attention.

Some of these incredibly false assertions are:

> "all natural chemicals are ipso facto safe, because nature always makes an antidote and allows adaptation to protect living creatures," whereas all synthetic chemicals are conversely "poisonous"
>
> "DDT is a carcinogen and a mutagen"
>
> "cancer in man is predisposed in human tissues by the inconsiderate ingestion of pesticide-treated foods during his mother's pregnancy"
>
> "great areas of our arable soil and forest lands have already been irrevocably contaminated permanently with synthetic chemical poisons"
>
> "foods produced with the use of agricultural chemicals are inferior in nutritional quality"
>
> "where no chemicals are employed, nature will provide more effective and safer control of insects, diseases, weeds, and predators by exerting biological control"

Absurd as these statements are, it has become an astounding reality that a majority of the American people believe them, if we are to believe recent polls, which may well be false, also.

How is this possible in this day and, particularly, in this country, where our agricultural education and extension facilities are by far the best in the world? Our USDA and land-grant college system, now celebrating its centennial, stands witness to the tremendous flow of education, research, and published information which has emerged over the past one hundred years. How is it possible that so much of this essential and valuable information has not reached the majority of our people?

The answer is simple. We in agriculture have shrunk to a distinct minority during the past twenty-five years, and have become so by dint of our own competence. We have, in effect, researched ourselves into obscurity both politically and socially. In addition, we have engendered the ire of the urbanite by producing a so-called surplus of food and fiber, which, it is claimed, cost the urban taxpayer several billion dollars per year to support.

This gross misunderstanding, punctuated all too frequently by crisis and sensationalism, is the target at which we need to fire our arsenal of scientific truth. We have been so preoccupied in agricultural research, rural education, and extension that we have essentially ignored the urban peoples in our communications, and have been talking largely to ourselves.

We need now to tell the urban peoples in a thousand places and a thousand ways what scientific agriculture, including agricultural chemistry, has meant to their health, their welfare, and their standard of living. We should make it clear in schools, in service clubs, in church meetings, and in the hundreds of other groups to which our people attach themselves:

> that the entire cost of agricultural research by federal, state, and industry is less than the savings it brings in cost of food alone to the American people each year;
>
> that DDT alone has saved as many lives over the past fifteen years as all the wonder drugs combined;
>
> that insecticides have been credited with extending the prospective life span in at least one Asiatic country from thirty-two to forty-seven years;
>
> that our so-called surplus of food is really no more of a surplus than a healthy reserve in the bank which represents our margin of safety in a hungry and envious world, and which ensures a steady course of food and prices in place of wide seasonal fluctuations which could cost our consumers several fold the expense of support programs;
>
> that our cheap, nutritious and wide array of foodstuffs in America today costs us less in terms of hours of work than it did any other nation since time began;
>
> that agricultural science with all its disciplines working in close collaboration in the lab, in the college, in the field, in the factory, and out on the farm, has for the first time in man's long struggle against want procured the means to banish hunger from the earth in our time;
>
> that our knowledge and control of the chemistry and function of the pesticides and additives we use vastly exceed that of the natural compounds which invariably contaminate our food supply when it is unprotected;

> that based on this cheap, safe, prolific, and varied food supply is man's ability to turn his surplus time and energy into research, education, and culture, for by DeGraff's law the progress of a people is inversely proportional to the time and effort required to produce the necessaries of existence.

This is what we must tell our urban peoples in all its brilliance and luster, and then rely upon their good sense to decide whether the alarums of the Longgoods, the Bicknells, and the Carsons are valid or whether the patient researches of scientific agriculture published from thousands of laboratories over the past century can be reliably accepted with confidence. One would think that the munificence around us would be evidence enough to decide. The mere fact that it is not is a clear indication that we have failed to tell our people in understandable parlance what we have done and why we have done it. We need to tell our story with vividness and inspiration to catch the imagination of the urban masses.

Miss Rachel Carson has done it from the opposite side in her book *Silent Spring*, for she is a writer on biological subjects with an extraordinary, vivid touch and elegance of expression. She paints a nostalgic picture of Elysian life in an imaginary American village of former years, where all was in harmonious balance with Nature and happiness and contentment reigned interminably, until sickness, death, and corruption was spread over the face of the landscape in the form of insecticides and other agricultural chemicals.

But the picture she paints is illusory, and she as a biologist must know that the rural Utopia she describes was rudely punctuated by a longevity among its residents of perhaps thirty-five years, by an infant mortality of upwards of twenty children dead by the age of five of every 100 born, by mothers dead in their twenties from childbed fever and tuberculosis, by frequent famines crushing the isolated peoples through long dark, frozen winters following the failure of a basic crop the previous summer, by vermin and filth infesting their homes, their stored foods and their bodies, both inside and out.

Surely she cannot be so naive as to contemplate turning our clocks back to the years when man was indeed immersed in Nature's balance and barely holding his own. Indeed, in many areas of the world, including some colonies in America, he failed to withstand the competition and ignominiously expired.

She ignores the fact that through the sciences she depreciates man can maintain himself today anywhere on earth. Within the past 100 years, man has emerged from a feeble creature, virtually at the mercy of Nature and his environment, to become the only being which can penetrate every corner of the planet, communicate instantly to anywhere on earth, produce all the food, fiber, and shelter he needs, wherever he may need it, change the topography of his lands, the sea and the universe and prepare his voyage through the very arch of heaven into space itself.

This is the stuff that science is made of, and man has learned to use it. He cannot now go back; he has crossed his Rubicon and must advance into the future armed with the reason and the tools of his sciences, and in so doing will doubtless have to contest the very laws and powers of Nature herself. He has done this already by expanding his numbers far beyond her tolerance and by interrupting her laws of inheritance and survival. Now, he must go all the way, for he cannot but partially contest Nature. He has chosen to lead the way; he must take the responsibility upon himself.

20

The Myth of the "Pesticide Menace"

EDWIN DIAMOND

Thanks to an emotional, alarmist book called *Silent Spring*, says this science writer, Americans mistakenly believe their world is being poisoned.

Thanks to a woman named Rachel Carson a big fuss has been stirred up to scare the American public out of its wits.

A year ago, in a book entitled *Silent Spring*, Miss Carson warned that pesticides were poisoning not only pests but birds and humans too. It was just what the public wanted to hear. No matter that Miss Carson's conclusions were preconceived; no matter that her arguments were more emotional than accurate. *Silent Spring* became a best seller and a conversational fad, and in Washington a congressional committee met to investigate the "pesticide menace."

Implied in this attack on pesticides are the much more serious charges that scientists are ignoring human values, experimenting for the sake of experiments, and upsetting the traditional "natural laws" and the so-called "balance of nature." Caught up in all the noise over *Silent Spring*'s revelations, we tend to forget, perhaps, that the lamentably widespread distrust of scientists and their works is anything but new.

When I was growing up in Chicago, I read, bug-eyed, a book that described the wholesale poisoning of the American public. A best seller of the time, the book was *100,000,000 Guinea Pigs*, and it recounted how an unholy trinity of government bureaucrats, avaricious businessmen and

Saturday Evening Post 236 (28 September 1963), 18 ff.

mad scientists had turned American consumers into laboratory test animals. I recall most vividly the danger ascribed to a certain toothpaste, which, if used in sufficient quantity, could cause a horrible death.

Today, a generation later, the American population has changed in many respects; for one, there are almost 190 million of us now instead of the 125 million around at the time the guinea-pigs book was written. In other respects, however, Americans are not much different. They still love to buy a book like Miss Carson's *Silent Spring* to read about their imminent death.

An indictment of the use of pesticides on farms, in forests and in suburban backyards, *Silent Spring* might just as easily have been called *190,000,000 Guinea Pigs*. For in *Silent Spring* I met again the old victims of my childhood dressed in more graceful prose. "As matters stand now," Miss Carson wrote, "we are in little better position than the guests of the Borgias." She conjured up an apocalyptic vision of a "silent spring," a time when plants, birds, animals—even humans—poisoned by DDT and other man-made chemicals sickened and died.

In June, 1962 *The New Yorker* serialized portions of *Silent Spring*. Newspapers carried stories about the book *before* its publication the following September. With this running start *Silent Spring* landed on the best-seller lists in a few weeks. The Book-of-the-Month Club offered it as the October selection. *CBS Reports* did two TV shows about it, and President Kennedy was questioned on pesticides at his news conferences. The ultimate accolade came when Miss Carson appeared before a Senate committee investigating pesticides. One of the Senators asked her for her autograph.

I have heard several theories to account for the vast stir that *Silent Spring* has created. First, there is Miss Carson's reputation and literary style. A quiet-spoken, retiring single woman, Miss Carson for sixteen years was employed as a biologist and later as editor-in-chief in the Bureau of Fisheries and the U.S. Fish and Wildlife Service. In 1951 she published the evocative and widely praised *The Edge of the Sea*. While *Silent Spring*—with its impassioned listing of case after case of sick sheep, sterile robins and dead fish—has little of the beauty of the earlier Carson books, it nevertheless has a huge expository gloss. As one of her critics said, "She's an alarmist and a sensationalist—and she's done it beautifully."

Second, there was the timing of the book coming as it did soon after the thalidomide drug tragedy had stirred Europe and the United States. Though *Silent Spring* does not deal with the licensing and marketing of new drugs, Miss Carson herself suggested in a newspaper interview at the time,

"It is all of a piece, thalidomide and pesticides. They represent our willingness to rush ahead and use something new without knowing what the results are going to be."

Third, there is the attention-getting quality inherent in any exaggeration. Rereading the reviews of *Silent Spring* not long ago, I found this echoed in such disparate journals as *The New York Times* ("She tries to scare the living daylights out of us and in large measure succeeds.") and the magazine *Scientific American* (". . . what I interpret as bias and oversimplification may be just what it takes to write a best seller").

Undoubtedly the noisy year that has followed publication of *Silent Spring* is as much a result of Miss Carson's alarmist approach as it is of her own literary reputation and the book's fortuitous—for sales—timing. But, at the risk of being charged with practicing psychology without a license, I'd like to suggest that there is another and less well understood reason for the popular reception given *Silent Spring* this past year.

Silent Spring, it seems to me, stirs the latent demons of paranoia that many men and women must fight down all through their lives. At one time or another all of us have been affected by the feeling that some wicked "they" were out to get "us." In recent years the paranoids among us could be observed in the ranks of such cultists as the antifluoridation leaguers, the organic-garden faddists and other beyond-the-fringe groups. And who are the "they" intent upon poisoning or tricking "us"? In the rough handbills passed out on street corners by the antifluoridationists, the plotters turn out to be Communists—scientists and dentists who want to soften, literally, the brains of the American citizenry to prepare them for Russian takeover by adding an insidious chemical to the drinking water.

In *Silent Spring* the villains aren't much more subtle. Miss Carson's "they" turn out to be the same tired stereotypes of *100,000,000 Guinea Pigs*. This is "an era dominated by industry, in which the right to make a dollar at whatever cost is seldom challenged," Miss Carson writes by way of explaining why the aerosol hiss of doom will continue unabated. Moreover, the greed of the businessman extends to the scientists; they, also, are venal. "The major chemical companies are pouring money into the universities to support research on insecticides," *Silent Spring* reveals darkly. "This creates attractive fellowships for graduate students and attractive staff positions. . . . This situation also explains the otherwise mystifying fact that certain outstanding entomologists are among the leading advocates of chemical control." That, of course, also explains away possible criticism in advance: Anyone who questions *Silent Spring* is obviously on Monsanto's or Shell's payroll. What it doesn't explain is why an industrialist or a scientist, no matter how grasping, would poison our food and water—the same food and water he himself eats and drinks.

But what about the accuracy of *Silent Spring*? Miss Carson's supporters frequently argue that, even though her book is slanted, she "got her basic facts right." Praising a science writer for getting the facts right, I would say, is like applauding a musician because he keeps time well. Equally important, however, are the facts that a writer leaves out and the half-facts or non-facts that are offered instead. *Silent Spring*, to take a hyperbolic example, speaks of the fall of "chemical death rain." This is vivid, but is it a "fact"? As Professor I. L. Baldwin of the University of Wisconsin noted in his review of *Silent Spring* in the authoritative journal *Science*, "Many may be led to believe that, just as rain falls on our land so is all of our land sprayed with pesticides. Actually, less than five percent of all the area of the United States is annually treated with insecticides."

Then too, there is the "fact," gravely stated by *Silent Spring*, that "for the first time in the history of the world, every human being is now subjected to contact with dangerous chemicals from the moment of conception until death." Assuming for a moment that this is true, what does it mean? In May of this year a nine-member panel of the President's Science Advisory Committee noted that deaths from the misuse of pesticides have numbered about 150 throughout the United States each year. To put this figure in perspective, consider these figures: The annual death toll from accidents involving aspirin is about 200 and from bee stings—yes *bee* stings—is about 150. No one, however, has seriously proposed eliminating the use of aspirin or exterminating all bees. Nor has anyone, with the possible exception of Miss Carson, proposed to abolish pesticides. As the panel put it, the more reasonable goal is to achieve "more judicious use of pesticides . . . to minimize risks."

Another of *Silent Spring*'s facts concerns the "many cats [that] are reported to have died" in western and central Java in the course of an anti-malarial program carried out by the World Health Organization. But we are told absolutely nothing about the cat's owners, the numberless Javanese men, women, and children who had previously suffered and died of malaria. Nor are we told anything about the fate of life—human life, not a cat's life—where there are no agricultural sprays or other modern food growing techniques. We hear nothing, for example, of the 10,000 people throughout the world who die of malnutrition or starvation every day. Nor do we read of the 1.5 billion people—more than half of the world's population—who live in perpetual hunger.

I mourn for the dead cats of Java and for the silent birds of the United States. I understand that the spraying of weed killer along roadsides also destroys some shelter for wildlife and therefore upsets the "balance of nature" so mystically evoked in *Silent Spring*. But is man to refrain from disturbing certain circumstances in nature that if kept in "balance" may balance him right out of existence? Science has been unable to find any such thing as a "balance" in nature, delicately turned and hovering around some fine ecological point. Nature has been altered by man ever since he first stood upright. If DDT kills some cats but saves many humans, if weed killer destroys a pocket of wildlife shelter but increases highway safety, so much the better.

Silent Spring, of course, is solicitous of humans, when the material suits its point of view. Miss Carson is particularly intent upon establishing a link between chemical sprays and a variety of diseases including cancer and mental disorders. She acknowledges that "it is admittedly difficult, in dealing with human beings rather than laboratory animals, to 'prove' that a cause A produces effect B. . . ." But this difficulty doesn't stop Miss Carson from attempting the same "proof" in the cause of hepatitis, an inflammation of the liver. First she cites the increased use of DDT, a chlorinated hydrocarbon, over the past two decades; then she notes that such chemicals can cause damage to the human liver; finally, she cites "the sharp rise in hepatitis that began during the 1950s and is continuing a fluctuating climb." Put them all together, she tells us, and "plain common sense" suggests that there is a relationship between the increase in liver ailments and the DDT spraying. Any number can play in the game of *post hoc, ergo propter hoc* reasoning: Nuclear testing also increased during the 1950s, so did television viewing, but there is no need to play the game at all, for many cases of hepatitis, as at least one critic of *Silent Spring* has pointed out, can be traced to infectious sources such as unsterilized hypodermic needles and water polluted by sewage.

What, finally, is *Silent Spring*'s game? If we were to believe Miss Carson's own description of our time—an era where the right to make an irresponsible dollar is seldom challenged—then the answer would be an easy one. But I believe this description, like so much else in *Silent Spring*, is an extravagant one.

A more accurate description would be that this is an era of stereotyped thinking, scattershot charges, shrill voices, and double standards of behavior. "I may not approve of Miss Carson's methods," someone is likely to say, "but she gets things done." In my experience, the speaker is usually the same person who a decade ago was most shocked by the flagrant techniques employed in Sen. Joseph McCarthy's Great Communist Hunt. The record shows that the nation, once down from its McCarthyite orbit, was able to deal with subversion without dismantling its noble mansion of constitutional law and civil rights. Similarly, I think the pesticide "Problem" can be handled without going back to a dark age of plague and epidemic.

21

DDT: Its Days Are Numbered, Except Perhaps in Pepper Fields

ROBERT GILLETTE

It would be premature to write the obituary of DDT, a chemical whose persistence in the biosphere has been more than matched by the perseverance of the pesticide industry and the Agriculture Department in sustaining its use under ten years of fire from environmentalists. But if DDT is not actually dead yet, it is clearly in its twilight days.

In retrospect, the turning point—and the beginning of the climactic chapter in DDT's turbulent history—was the Nixon Administration's decision in 1970 to shift federal pesticide authority from the Agriculture Department to the newly created Environmental Protection Agency. Last week's order by William Ruckelshaus, administrator of the EPA, to ban virtually all remaining domestic uses of DDT at the end of the year, closed that chapter. Now, for its epilogue, the struggle moves back to the federal courts, where lawyers for both the EPA and environmental groups seem confident that DDT's death warrant will be upheld.

Ruckelshaus' decision marked the end of a long series of administrative appeal proceedings open to the industry under the Federal Insecticide, Fungicide, and Rodenticide Act. These opportunities for second thoughts on the EPA's part spanned nearly eighteen months, dating from January

Science 176 (23 June 1972), 1313–34.

1971, when environmental groups won a federal appeals court order asking the EPA to cancel the government's formal approval of DDT as an "economic poison." The EPA complied, and the industry immediately requested its full due under the law—mainly in the form of a study of DDT's benefits and hazards by a panel of scientists nominated by the National Academy of Sciences (and selected by the EPA) and in the form of a quasi-judicial public hearing (*Science*, 10 December 1971). In the end, the NAS panel called for virtual elimination of DDT, but the hearing was another matter. A sometimes stormy affair, it lasted seven months, then finally brought a ruling from federal hearing examiner Edmund Sweeney that DDT's benefits tended to outweigh its risks and that certain "essential" uses should be retained. Among those uses was the protection of cotton, which accounted for 86 percent of the twelve to fourteen million pounds of DDT sprayed in the United States in 1970.

Sweeney's ruling was not binding on the EPA, however, and, while Ruckelshaus avoided explicitly saying so, he effectively reversed it. In the text of a forty-page decision, Ruckelshaus wrote, "The evidence of record showing storage [of DDT and its metabolites] in man and magnification in the food chain is a warning to the prudent that man may be exposing himself to a substance that may ultimately have a serious effect on his health."

That opinion coincided closely with the view of the NAS panel. It also paralleled the sentiments of a 1963 report by the President's Science Advisory Committee advocating the eventual "elimination" of persistent pesticides. And it is worth noting that still another major pesticide study group, the so-called Mrak Commission appointed by former Secretary of Health, Education, and Welfare Robert Finch, urged in December 1969 that the federal government "eliminate within two years all uses of DDT and DDD in the United States," except those uses for which no substitutes are available. That is precisely what the EPA has now decided to do—and only one year behind the Mrak timetable.

The banning of DDT was an act of political courage that went considerably further than a number of EPA staff members were willing to predict late last year, and it certainly went further than the federal courts were able to persuade the Agriculture Department to go while it still held sway over pesticides. Most directly, Ruckelshaus' decision dealt a blow to Representative Jamie L. Whitten (D-Miss.), a cotton state Congressman whose appropriations subcommittee controls funds for the EPA. Whitten's ardent support for the agricultural chemical industry in general and for DDT in particular is set forth in his 1966 book *That We May Live*, a rejoinder to the late Rachel Carson's *Silent Spring*.

The ban imposed by the EPA is not quite absolute, however. It does not affect the annual exportation of some thirty million pounds of DDT, nor does it prohibit government agencies from using DDT in public health emergencies. Moreover, the EPA left the door ajar for minor applications of the pesticide to three crops—onions in the Pacific Northwest, sweet potatoes in storage, and green peppers grown on the Delmarva Peninsula along the Chesapeake Bay. (Unless growers or the industry present some compelling new evidence to support the use of DDT on these crops within thirty days, the ban will be extended to include them as well.)

In the case of the green peppers, the loophole Ruckelshaus left attests more to the influence of an old Washington law firm than to the voraciousness of the corn borers that allegedly threaten to devour the Delmarva's peppers in the absence of DDT. The exemption came about at the behest of Henry P. Cannon and Sons, the peninsula's leading pepper processor, which hired the Washington firm of Covington and Burling to plead its case during one day of the seven-month hearing. Oddly enough, the pesticide industry itself raised no objection to the cancellation of DDT's registration for use on peppers, and, as Ruckelshaus candidly admitted in his decision, his own staff had advised against granting exemptions to any crops. "All this yelling about DDT is totally unfounded," Henry Cannon said by telephone from Bridgeville, Del. "Who's it ever hurt?"

If Cannon continues to win his way, he—or the local growers who contract to him—may be the last to use DDT in the United States. While the 13,500 pounds of DDT they apply to the peninsula's land each year may be inconsequential on a national scale, local environmentalists argue that it is a significant input to the Chesapeake Bay and to the bay's vulnerable population of ospreys.

Still, environmentalists have professed themselves generally pleased at the EPA's action. "We won more than 99 percent of what we wanted," said William Butler, a Washington attorney for the Environmental Defense Fund (EDF), the group most closely identified with opposition to DDT.

Indeed, the victory was especially sweet for the EDF, which owes its existence to DDT. The organization began with a small and unusual nucleus of New York scientists and lawyers who banded together in 1966 to protest the use of DDT for mosquito control in Suffolk County, Long Island. From this community squabble, the EDF has since grown to the stature of a national organization, with 32,000 paying subscribers; a pool of 700 scientists on call as expert witnesses; and offices in New York, Washington, and Berkeley. The EDF is currently a party to forty-odd cases running the gamut from air pollution to water resource litigation.

As the organization grew, it escalated the fight over DDT to the federal level. In October 1969, the EDF, representing itself and four other groups, petitioned then Secretary of Agriculture Clifford Hardin to halt interstate sales of DDT. Under threat of court action, Hardin did eliminate home use of DDT and some fifty other minor applications. When federal pesticide authority changed hands to the EPA, the EDF redirected its petition for a complete domestic ban—this time successfully.

During the public hearings that ensued, environmental groups coalesced with the EDF and joined the EPA's Pesticides Office as an equal partner in defending the proposed ban; the Agriculture Department, as if to substantiate Ralph Nader's characterization of it as the "Department of Agribusiness," joined the case on the side of the industry.

With appeals to federal agencies now exhausted by the industry, the action has shifted to two federal courts. One is the Fifth Circuit Court of Appeals in New Orleans, where the industry is now seeking to have the EPA's ban set aside by a panel of judges it apparently hopes will be more sympathetic to the cotton industry than to federal agencies and environmental groups. With the opposite strategy in mind, the EDF is seeking to make the ban immediate—and to keep the case in Washington—by a motion pending before the Court of Appeals of the District of Columbia. Few observers, however, see much chance of either court reversing the EPA, particularly since the ban does not affect public health applications of DDT—which have dwindled almost to the point of nonexistence in the United States anyway.

It may still be, of course, that DDT's proponents fear that its use in the United States is only the first of the dominoes to fall and that a hasty worldwide ban may follow. Such fears may be exaggerated, though, if the views of EDF's Butler are any indication. He says that he personally accepts the World Health Organization's argument that DDT is still essential for controlling disease in less developed nations. "What we hope will happen," Butler says, "is that other nations will begin to question for themselves the advisability of using DDT in agriculture. We think that a combination of less persistent pesticides and proper crop management can be more economical than DDT."

PART 5

DDT AND MALARIA

The recent debates about DDT and malaria, though they focus on DDT, have as background almost a century of research on tropical diseases, particularly malaria, that established how difficult effective control could be. In the 1920s the Rockefeller Foundation supported research that worked out the connections among disease organisms, mosquitoes, and humans, and many countries used these methods as the basis for control until the introduction of DDT. They aimed not at killing the disease organism but breaking the chain of infection by killing the mosquitoes that carried it or keeping them away from humans. These often required labor-intensive controls such as drying up breeding places near people's homes by filling in puddles and removing tin cans, old tires, and anything else that held a little water. Other methods included killing larvae by introducing chemicals into the water they lived in, introducing fish that ate these "wrigglers," or putting nets and screens in houses. In a few places mosquitoes were vulnerable to eradication programs, but most were not. Upon its introduction DDT swept away these methods, for it was inexpensive, deadly, and far easier to use and keep in use.

After 1970 it fell out of favor, though not for any simple reason. The American ban had an effect, but so did rising resistance in insect populations, competition from newer chemicals, and changes in the places and ways people lived that made control with chemicals harder. In the 1990s rising rates of malaria made tropical nations reluctant to support a proposed treaty outlawing DDT; some wanted it as a reserve weapon or for immediate use. That situation brought a new round of polemics in the United States, as anti-environmentalists argued that it showed Rachel Carson had been wrong, tragically mistaken, or simply unconcerned about the impact of her arguments. Environmentalists defended Carson, the ban, and the movement.

These selections suggest the range of views. In "Intended Consequences," Thomas Sowell, an economist with the Hoover Institution, presented the case against Rachel Carson and the environmental movement, arguing that her callousness resulted in the deaths of millions in the Third World. In "Re-reading *Silent Spring*," Thomas R. Hawkins reflected on her work and legacy forty years after her death and came to very different conclusions. In "If Malaria's the Problem, DDT's Not the Only Answer," May Berenbaum, an economic entomologist, took another tack entirely, seeing DDT as a tool rather than a solution, something not good or bad in itself but useful in some situations and not in others. These authors have perspectives as diverse as their conclusions. Sowell speaks of public policy in terms of economic development, Hawkins of the personal effect of reading Rachel Carson's work, and Berenbaum comes at the question from the standpoint of economic entomology and the perspective of a scientist trying to form public policy.

All three pieces speak to current ideas and concerns but also to ideas and values that go back to the earliest readings. Both Sowell and Hawkins, for example, approach the issue of DDT and the reputation of Rachel Carson with the hindsight of forty years of environmental action, while Berenbaum recalls, though she does not recite, the long history of malaria control programs. In that sense all are part of the present and address our current situation. Sowell, though, also speaks to a long tradition of concern with national economic development, Hawkins to a vibrant tradition of nature preservation, and Berenbaum to the economic entomologists' necessary involvement in politics and public relations. Problems come and go, but underlying attitudes just take new forms.

22

Intended Consequences

THOMAS SOWELL

Over the years, the phrase "unintended consequences" has come up with increasing frequency, as more and more wonderful-sounding ideas have led to disastrous results. By now, you might think that people with wonderful-sounding ideas would start to question what the consequences would turn out to be—and would devote as much time to discovering those consequences as to getting their ideas accepted and turned into laws and policies. But that seldom, if ever, happens.

Why doesn't it? Because a lot depends on what it is you are trying to accomplish. If your purpose is to achieve the heady feeling of being one of the moral elite, then that can be accomplished without the long and tedious work of following up on results.

The worldwide crusade to ban the pesticide DDT is a classic example. This crusade was begun by the much revered Rachel Carson, whose best-selling book *Silent Spring* was based on the premise that DDT's adverse effects on the eggs of song birds would end up wiping out these species. After that, springtime would no longer be marked by birds singing; hence the silent spring.

Rachel Carson and the environmentalists she inspired have succeeded in getting DDT banned in country after country, for which they have received

htpp://www.jewishworldreview.com

the accolades of many, not least their own accolades. But, in terms of the actual consequences of that crusade, there has not been a mass murderer executed in the past half-century who has been responsible for as many deaths of human beings as the sainted Rachel Carson. The banning of DDT has led to a huge resurgence of malaria in the Third World, with deaths rising into the millions.

This pioneer of the environmental movement has not been judged by such consequences, but by the inspiring goals and political success of the movement she spawned. Still less are the environmentalists held responsible for the blackouts plaguing California in the past year or the more frequent blackouts and more disastrous economic consequences that can be expected in the years ahead, despite the key role of environmental extremists in preventing power plants from being built.

The greens have likewise obstructed access to the fuels needed to generate electricity, run automobiles and trucks, and perform innumerable other tasks in the economy. Nationwide, the greens have been so successful in preventing oil refineries from being built that the last one constructed anywhere in the United States was built during the Ford administration. But environmentalists are seldom mentioned among the reasons for today's short supplies of oil and the resulting skyrocketing prices of gasoline.

Advocates of rent control are not judged by the housing shortages that invariably follow, but by their professed desire to promote "affordable housing" for all. Nor are those who have promoted price controls on food in various countries being judged by the hunger, malnutrition or even starvation that have followed. They are judged by their laudable goal of seeking to make food affordable by the poor—even if the poor end up with less food than before.

Some try to argue against the evidence for these and other counterproductive consequences of high-sounding policies. But what is crucial is that those who advocated such policies usually never bothered to seek evidence on their own—and have resented the evidence presented by others. In short, what they advocated had the intended consequences for themselves—making them feel good—and there was far less interest in the unintended consequences for others.

Even before the rise of today's many social activist movements, T. S. Eliot understood such people and their priorities. Writing in 1950, he said: "Half the harm that is done in this world is due to people who want to feel important. They don't mean to do harm—but the harm does not interest them.

Or they do not see it, or they justify it because they are absorbed in the endless struggle to think well of themselves."

There is little hope of changing such people. But what the rest of us can do is stop gullibly accepting their ego trips as idealistic efforts for others. Above all, we need to stop letting them morally intimidate us into silence about the actual consequences of their crusades. The time is long overdue for us to insist that they put up or shut up, in terms of hard evidence about results, rather than the pious hopes that make them feel so good.

23

Re-reading *Silent Spring*

THOMAS R. HAWKINS

In 1962, *The New Yorker* magazine serialized substantial portions of the book manuscript *Silent Spring*, which critically examined the use of pesticides in controlling insects and the effects of these chemicals on the broad spectrum of life, including wildlife and human health. The author was Rachel Carson, a fifty-four-year-old former employee of the U.S. Fish and Wildlife Service. Both her placement of the serialized portions of the book in *The New Yorker* and the controversial issues Carson raised won a wide audience for her work, an audience which grew and launched Carson into national visibility when the book was published by Houghton Mifflin later in 1962.

In seventeen concise chapters, many of which can stand alone as essays, Carson develops a deceptively simple premise: the use and overuse of synthetic chemicals to control insect pests introduces these chemicals into the air, water, and soil and into the food chain where they poison animals and humans, and disrupt the many intricate interdependencies that make up the delicate natural order. In the concluding paragraph of the book, Carson said:

Environmental Health Perspectives 102 (Nos. 6–7, June–July 1994), 536–37.

> The "control of nature" is a phrase conceived in arrogance, born of the Neanderthal age of biology and philosophy, when it was supposed that nature exists for the convenience of man. The concepts and practices of applied entomology for the most part date from that Stone Age of science. It is our alarming misfortune that so primitive a science has armed itself with the most modern and terrible weapons, and that in turning them against the insects it has also turned them against the earth.

Reflecting on the impact of the book when it was first published, David P. Rall, director of NIEHS [National Institute of Environmental Health Sciences] from 1971 until his retirement in 1990, and the founding director of the National Toxicology Program, said, "In many ways, *Silent Spring* was the beginning of the environmental movement. It was the first serious look at the persistence of environmental chemicals, and one of the most important books of the 1960s."

The "list of principal sources" at the end of the book, an expansive bibliography that spares the reader footnotes in the text, runs a full fifty-five pages. "Rachel Carson brings real insight to her subject," Rall said. "She does this partly by pulling together material from disparate sources, and also through her elegant writing style, which makes it easy to get educated on this subject."

The value of reading or re-reading *Silent Spring* today resides in Rall's observation; it remains among the most concise and best-written overviews on the subject of pesticides, eerily fresh after nearly a third of a century, with many of the topics still emerging as issues in science, biology, ecology, and public health. Carson didn't just sell millions of books and raise a stir among chemical manufacturers and politicians, she shaped perceptions in a lasting way.

"Rachel Carson did not just affect my career," said Lynn Goldman, assistant administrator of the Office of Prevention, Pesticides and Toxic Substances, of EPA. "Carson affected an entire generation in their understanding of the environment." Goldman recalls, "I grew up in Galveston, Texas, and when the pelicans began to disappear from Galveston Bay [because of the use of DDT], I felt that loss very strongly. Then in the early 1980s, after DDT had been banned, it was wonderful when the pelicans returned. This was really quite a lesson. If we see these reproductive effects in birds, we might expect to see evidence of Rachel Carson's hypothesis that there are health effects in humans, especially now that we have the tools to better understand the human effects."

Goldman continued, "I'm surprised that I still hear from people who think DDT never harmed the environment. We have a long way to go in understanding pesticides as they are related to the environment. It is amazing how much Rachel Carson understood, even in the early 1960s, about biodiversity and ecosystems and the relationship between pesticides, the environment, and health."

It is not only environmental professionals who have this sense of having been moved and permanently affected by *Silent Spring*. The book seemed to impact almost everyone who read it and eventually many who hadn't. Said one bookseller who recently sold out of used copies of the book, "I read *Silent Spring* when I was in my 20s, and for a year or two I thought I was the only environmentalist in the world." Rachel Carson's work still has the power to awaken a profound sense of connection between human beings and the rest of the natural world.

Despite the deep chords of agreement that Carson struck among many of her readers, or perhaps because of them, there were many who disagreed with her, and many of those who disagreed were aggressive and vocal. Some, in the chemical industry, for example, launched counter-attacks: first against Carson's professional credentials and scientific arguments and, in some instances, personal attacks. The mudslinging often served only to further elevate Carson's reputation among her admirers.

An obituary of Carson in the 24 April 1964 issue of *Time* probably reflected the skepticism of the magazine's mass audience when it stated:

> To its author [*Silent Spring*] was more than a book; it became a crusade. And, despite her scientific training, she rejected facts that weakened her case, while using almost any material, regardless of authenticity, that seemed to support her thesis. Her critics, who included many eminent scientists, objected that the book's exaggerations and emotional tone played on the vague fears of city dwellers, the bulk of the U.S. population, who have little contact with uncontrolled nature and do not know how unpleasantly hostile it generally is. Many passages mentioned cancer, whose cause is still mysterious. Who knows? suggested the book. Could one cause of the disease be pesticides?

The *Time* article also attributed continuing repercussions from widespread publication of the book: "Laws were proposed on local, state and federal levels to put rigid restrictions on the use of pesticides. Some of them were so sweeping that if they had been passed and enforced, they might very well have caused serious harm. In advanced modern societies, agriculture and public health can no longer manage without chemical pesticides." The irony of this statement is obvious in light of recent scientific evidence of the dangers of pesticides and the series of environmental regulations that continue to limit their use.

At the time, however, criticisms were heaped on the book in an effort to discredit Carson's premise. Robert White-Stevens of American Cyanamid, a spokesperson for the chemical industry, was quoted In the 27 April 1964 issue of *Newsweek*, saying, "The major claims . . . are gross distortions of the actual facts, completely unsupported by scientific, experimental evidence, and general practical experience in the field."

The *Newsweek* article attempted to downplay the impact of the book by questioning Carson's role in precipitating legal and regulatory changes:

> It is difficult to isolate the effect of the book, for the case against indiscriminate use of chemicals was already being aired before it was published. . . . A Federal study of pesticides was under way, and when the report was published in 1963 it stated that chemicals were potentially very dangerous, and advised that controls must be strict and well enforced. It also pointed out the side Miss Carson chose to ignore: that chemicals are in large part responsible for the increase in U.S. agricultural productivity and have helped control such diseases as malaria.

In rebuttal to these criticisms, Stewart L. Udall, then Secretary of the Interior, defended Carson in a *Saturday Review* article in the 16 May 1964 issue. Said Udall, "*Silent Spring* was called a one-sided book. And so it was. She [Carson] did not pause to state the case for the use of poisons on pests, for her antagonists were riding roughshod over the landscape. They had not bothered to state the case for nature. The engines of industry were in action; the benefits of pest control were known—and the case for caution needed dramatic statement if alternatives to misuse were to be pursued."

More than twenty years later, Carson was still catching literary and scientific heat, posthumously. Author and social commentator Edith Efron, in her expansive 1984 critique of environmental sciences, *The Apocalyptics*, rated Carson as "the first apocalyptic of national importance." Efron's thesis is that the environmental movement has been fraught from the beginning with emotionally and politically skewed thinking that muddles clear scientific reasoning. Efron asserted that Carson promoted the idea that there are relatively few natural carcinogens—arsenic, a few kinds of radioactive rocks, and sunlight—whereas man-made chemicals are the source of an increased incidence of cancer. In response to Carson's statement that humans created their own cancerous universe because only humans can create cancer-causing substances, Efron concludes:

> It is entirely apparent that Carson's analysis of the carcinogen problem is the very analysis that now prevails among American regulators; and it is also apparent that among some at the National Cancer Institute, the concept of cancer as a political disease requiring political solution had been fully crystallized at least two decades ago. But more important yet, Carson's analysis tells us that the apocalyptic approach to cancer rests, fundamentally, on the "axiom" of a largely benevolent nature—a vision of a largely noncarcinogenic Garden of Eden now defiled by the sins of pride and greed.

That Carson's concern about synthetic chemicals hinges in some fundamental way on the significance of natural carcinogens is considered by some as a diversionary argument. Clearly, the discussion of carcinogenicity proceeds along an extensive continuum, with much debate and discussion at every point along the many gradations of opinion. Nonetheless, even those who disagree with Carson recognize the lasting influence of *Silent Spring.* "The influence of Carson on our era can hardly be overstated," Efron said, " . . . the Toxic Substances Control Act under which we live today is a monument to her thought."

The persistence of *Silent Spring* as an environmental touchstone has been its accuracy in predicting emergent issues. Early response to Carson's book centered on concerns about effects on wildlife and, in human terms, on environmentally mediated cancer. But Carson specifically mentions human reproductive effects as a possible disease endpoint for environmental exposures, an area of concern that is just now receiving greater scientific attention. "Throughout her book, Rachel Carson reported that pesticides were capable of affecting fertility and even discussed research where animals deliberately exposed to pesticides in the laboratory never reached sexual maturity," said Theodore Colborn, senior scientist with the World Wildlife Fund. "At the time she wrote the book, eggshell thinning and outright mortality among wildlife were common—the results of heavy, unregulated use of pesticides. The high-dose exposure masked the less visible effects that lead to loss of fertility and other physiological functions," Colborn said.

Colborn notes that the human health effects took longer to become apparent. "Among humans, a long-lived species, the evidence of transgenerational effects was only beginning to be played out in the individuals exposed to pesticides in the womb—individuals whose loss of function would not be expressed for another ten or twenty years as they matured. Carson's book was a documentary on what was evident at that time—cancer and acute toxicity, effects expressed in directly exposed individuals—which preoccupied the minds of millions around the world after the detonation of the first atomic bomb. Unfortunately, those charged with protecting human and wildlife health focused largely on cancer to determine the safety of man-made products. As a result, the delayed, long-term, adverse health effects of pesticides that lead to loss of species were overlooked."

Silent Spring, both as a work of literature and a call for social and scientific scrutiny of the use of pesticides, shows every evidence of enduring into the millennium because Carson presented a premise on the relationship between humans, the use of chemicals, and the environment that has been borne out by science. Despite the emotional dimensions of the subject, proponents of the work credit Carson for adhering to rigorous standards of evidence and relentlessly researching her subject. To many, although her tone was modest and workmanlike, her insights and intuitions were inspired. Whether one agrees with Carson's premise or not, *Silent Spring* stands among the best read and most revered books on science addressed to a general audience. In the final analysis, even Efron labeled the work "a living classic."

24

If Malaria's the Problem, DDT's Not the Only Answer

MAY BERENBAUM

In the pantheon of poisons, DDT occupies a special place. It's the only pesticide celebrated with a Nobel Prize: Swiss chemist Paul Mueller won in 1948 for having discovered its insecticidal properties. But it's also the only pesticide condemned in pop song lyrics—Joni Mitchell's famous "Hey, farmer, farmer put away your DDT now"—for damaging the environment. Banned in the United States more than thirty years ago, it remains America's best known toxic substance. Like some sort of rap star, it's known just by its initials; it's the Notorious B.I.G. of pesticides.

Now DDT is making headlines again. Many African governments are calling for access to the pesticide, believing that it's their best hope against malaria, a disease that infects more than 300 million people worldwide a year and kills at least three million, a large proportion of them children. And this has raised a controversy of Solomonic dimensions, pitting environmentalists against advocates of DDT use.

The dispute between them centers on whether the potential benefits of reducing malaria transmission outweigh the potential risks to the environment. But the problem isn't that simple. This is a dispute in which science should play a significant role, but what science tells us is that DDT is nei-

Washington Post, Outlook Section (5 June 2005), B3.

ther the ultimate pesticide nor the ultimate poison, and that the lessons of the past are being ignored in today's discussion.

The United Nations Environment Program has identified DDT as a persistent organic pollutant that can cause environmental harm and lists it as one of a "dirty dozen" whose use is scheduled for worldwide reduction or elimination. But some DDT advocates have resorted to anti-environmentalist drama to make their case for its use in Africa.

They have accused environmental activists of having "blood on their hands" and causing more than 50 million "needless deaths" by enforcing DDT bans in developing nations. In his best-selling anti-environmentalist novel *State of Fear*, Michael Crichton writes that a ban on using DDT to control malaria "has killed more people than Hitler."

Such statements make good copy, but in reality, chemicals do not wear white hats or black hats, and scientists know that there really are no miracles.

Malaria is caused by a protozoan parasite that is transmitted by mosquitoes. For decades, there have been two major strategies for curbing the disease: killing the infectious agent or killing the carrier. Reliably killing the protozoan has proved difficult; many older drugs are no longer effective, new ones are prohibitively expensive, and delivering and administering drugs to the susceptible populace presents daunting challenges. Killing the carrier has long been an attractive alternative.

And DDT has been an astonishingly effective killer of mosquitoes. DDT (which stands for the far less catchy dichloro-diphenyl-trichloroethane) is a synthetic chemical that didn't exist anywhere on the planet until it was cooked up for no particular purpose in a German laboratory in 1874. Decades later, in 1939, Mueller pulled it off a shelf and tested it, along with many other synthetic substances, for its ability to kill insects. DDT distinguished itself both by its amazing efficacy and its breadth of action—by interfering with nervous system function, it proved deadly to almost anything with six, or even eight, legs. And it was dirt-cheap compared to other chemicals in use—it could be quickly and easily synthesized in chemical laboratories from inexpensive ingredients.

Soon after its insecticidal properties were discovered, DDT was put to use combating wartime insect-borne diseases that have bedeviled troops mobilized around the world for centuries. It stemmed a louse-borne typhus outbreak in Italy and prevented mosquito-borne diseases in the Pacific theater, including malaria and yellow fever, to almost miraculous effect. This military success emboldened governments around the world to use DDT after World War II to try to eradicate the longtime scourge of malaria. And in many parts of the world, malaria deaths dropped precipitously. This spectacular success is why many people are calling for the use of DDT specifically for malaria control.

At the same time that malaria deaths were dropping in some places, however, the environmental persistence of DDT was creating major problems for wildlife, as famously documented in Rachel Carson's classic 1962 book, *Silent Spring.* By 1972, the pesticide had become the "poster poison" for fat-soluble chemicals that accumulate in food chains and cause extensive collateral damage to wildlife (including charismatic predators such as songbirds and raptors), and a total ban on the use of DDT went into effect in the United States.

What people aren't remembering about the history of DDT is that, in many places, it failed to eradicate malaria not because of environmentalist restrictions on its use but because it simply stopped working. Insects have a phenomenal capacity to adapt to new poisons; anything that kills a large proportion of a population ends up changing the insects' genetic composition so as to favor those few individuals that manage to survive due to random mutation. In the continued presence of the insecticide, susceptible populations can be rapidly replaced by resistant ones. Though widespread use of DDT didn't begin until WWII, there were resistant houseflies in Europe by 1947, and by 1949, DDT-resistant mosquitoes were documented on two continents.

By 1972, when the U.S. DDT ban went into effect, nineteen species of mosquitoes capable of transmitting malaria, including some in Africa, were resistant to DDT. Genes for DDT resistance can persist in populations for decades. Spraying DDT on the interior walls of houses—the form of chemical use advocated as the solution to Africa's malaria problem—led to the evolution of resistance forty years ago and will almost certainly lead to it again in many places unless resistance monitoring and management strategies are put into place.

In fact, pockets of resistance to DDT in some mosquito species in Africa are already well documented. There are strains of mosquitoes that can metabolize DDT into harmless byproducts and mosquitoes whose nervous systems are immune to DDT. There are even mosquitoes who avoid the toxic effects of DDT by resting between meals not on the interior walls of houses, where chemicals are sprayed, but on the exterior walls, where they don't encounter the chemical at all.

The truth is that DDT is neither superhero nor supervillain—it's just a tool. And if entomologists have learned anything in the last half-century of dealing with the million-plus species of insects in the world, it's that there is no such thing as an all-purpose weapon when it comes to pest management. DDT may be useful in controlling malaria in some places in Africa, but it's essential to determine whether target populations are resistant; if they are, then no amount of DDT will be effective.

We have new means of determining whether populations are genetically prone to developing resistance. DDT advocates are right to suggest that DDT may be useful as a precision instrument under some circumstances, particularly considering that environmental contamination in Africa may be less of a problem than it has been in temperate ecosystems because the chemical can degrade faster due to higher temperatures, moisture levels and microbial activity. Moreover, resistance evolves due to random mutation, so there are, by chance, malaria-carrying mosquito species in Africa that remain susceptible to DDT despite more than two decades of exposure to the chemical.

But environmentalists are right to worry that the unwise use of DDT, particularly where it is likely to be ineffective, may cause environmental harm without any benefit. In 2000, I chaired a National Research Council committee that published a study titled "The Future Role of Pesticides in U.S. Agriculture." Our principal recommendation is germane to discussions of malaria management: "There is no justification for completely abandoning chemicals per se as components in the defensive toolbox used for managing pests. The committee recommends maintaining a diversity of tools for maximizing flexibility, precision, and stability of pest management."

Overselling a chemical's capacity to solve a problem can do irretrievable harm not only by raising false hopes but by delaying the use of more effective long-term methods. So let's drop the hyperbole and overblown rhetoric—it's not what Africa needs. What's needed is a recognition of the problem's complexity and a willingness to use every available weapon to fight disease in an informed and rational way.

Notes on Further Reading

Suggestions about background reading must be general, given the amount of material out there. Regard this as a set of places to start exploring, not a definitive guide.

On humans' relation to nature in our culture, two classic works provide deep background: Clarence Glacken's *Traces on the Rhodian Shore* (Berkeley: University of California Press, 1967) deals with ideas about nature in Western civilization from antiquity to the Enlightenment, and Roderick Nash's *Wilderness and the American Mind* (New Haven: Yale University Press, 4th edition, 2001) brings the story closer to home. Notes in each provide ample sources for further reading, and the later chapters in Nash's work have places to start reading about the environmental movement. A less historical but more philosophical approach can be found in Rene Dubos's *Mirage of Health* (Garden City: Doubleday, 1959), which argues, with information from medical history as well as medicine, that the "conquest of nature" must remain a dream.

On science, the authoritative source of knowledge about nature in our culture, Peter Bowler's *The Earth Encompassed* (New York: Norton, 1992) gives an overview of the development of the earth and biological sciences that grew from natural history. It too has suggestions for further reading. Work on popular understanding and amateur studies since the nineteenth century must start with Britain, which so strongly influenced America. See

David Allen's *The Naturalist in Great Britain* (London: Allen Lane, 1976) and Lynn Barber's *The Heyday of Natural History* (London: Jonathon Cape, 1980). For other perspectives see John MacKenzie, *Empire of Nature* (New York: Manchester University Press, 1988), and Harriet Ritvo, *Animal Estate* (Cambridge: Harvard University Press, 1987). Peter Schmitt, *Back to Nature* (New York: Oxford University Press, 1969), discusses the American movement of the late nineteenth century, and Jenny Price, *Flight Maps* (New York: Basic Books, 1999), has a set of perspectives and studies on the twentieth. On ecology and our view of nature, Donald Worster's *Nature's Economy* (San Francisco: Sierra Club, 1977) sketches in the basic story of ecology as a discipline and a way of thinking from a stance outside science; Robert McIntosh's *Background to Ecology* (New York: Cambridge University Press, 1985) approaches it from further in. On science, see also Sharon Kingsland, *Modeling Nature* (Chicago: University of Chicago Press, 1985), and Frank Golley, *History of the Ecosystem Concept in Ecology* (New Haven: Yale University Press, 1993). For wildlife and wildlife policy, the notes in my *Saving America's Wildlife* (Princeton: Princeton University Press, 1988) provide a guide. These are historians' approaches. For other, recent perspectives, look up the Society for Literature and the Environment.

On the main topic here, pesticides, James Whorton's *Before Silent Spring* (Princeton: Princeton University Press, 1974) deals with the early insecticides, and my own *DDT: Scientists, Citizens, and Public Policy* (Princeton: Princeton University Press, 1981) traces that story from its beginnings through the federal ban. Christopher Bosso adds the politics of pesticide lobbying and the rise of environmental politics in *Pesticides and Politics* (Pittsburgh: University of Pittsburgh Press, 1987), and Edmund Russell looks at the rhetoric of the "war against insects" in *War and Nature* (New York: Cambridge University Press, 2001). L. O. Howard's autobiography, *Fighting the Insects* (New York: Macmillan, 1933), and his *The Insect Menace* (New York: Century, 1931) provide contemporary evidence. For a researcher's view of the links between industry and research in the DDT years, see Robert van den Bosch, *The Pesticide Conspiracy* (Garden City: Doubleday, 1978).

Paul Brooks, Rachel Carson's editor at Houghton Mifflin, did a study of her literary work, *House of Life* (Boston: Houghton Mifflin, 1972), and environmental historian Linda Lear produced a full-scale scholarly biography, *Rachel Carson* (New York: Henry Holt, 1997). Several books have traced the controversy over *Silent Spring*, starting with Frank Graham, *Since Silent*

Spring (Boston: Houghton Mifflin, 1970). Others include H. Patricia Hynes, *The Recurring Silent Spring* (New York: Pergamon, 1989), a feminist analysis and an account of continuing environmental problems from chemical contamination; Mark Hamilton Lytle, *The Gentle Subversive* (New York: Oxford University Press, 2007), a biography in the context of American environmentalism; Alex MacGillivray, *Rachel Carson's Silent Spring* (New York: Barronn's, 2004), a short, popular treatment in the series "Words That Changed the World"; and Priscilla Coit Murphy, *What a Book Can Do: The Publication and Reception of Silent Spring* (Amherst: University of Massachusetts Press, 2005).

Opposition to Rachel Carson came in articles or reviews rather than books. The National Agricultural Chemicals Association, which led the initial charge, relied in particular on William J. Darby's "A Scientist Looks at *Silent Spring*," *Chemical and Engineering News*, 40 (1 October 1962), 60–63, also printed as a flyer by the American Chemical Society, and Ira L. Baldwin's "Chemicals and Pests," *Science*, 137 (28 September 1962), 1042–43. The Internet supplies an enormous, unsorted array of information ranging from useful to so far from reality it cannot even be called wrong. After 1972, discussions of Carson became entangled in successive major environmental issues, where many of the issues raised about DDT—our ability to control nature with technology, the unforeseen consequences of technologies, the workings of nature (did it have a "balance" and what did that mean?), and humans' relation to nature—returned in other forms. Places to start include Donella H. Meadows, Jørgen Randers, and Dennis Meadows, *The Limits to Growth* (New York: Universe, 1974), for the debate over the earth's limits on supporting human populations and the polemical literature around it, and E. O. Wilson, *The Diversity of Life* (Cambridge: Harvard University Press, 1992), on the impending loss of biodiversity and, most recently, global warming.

These controversies say something about the environmental movement that grew up in the wake of the controversy over *Silent Spring*, pesticides, and pollution, but that movement generated a scholarly literature almost from birth—useful for perspective on the various controversies and values. Places to start include Samuel P. Hays, *Beauty, Health, and Permanence* (New York: Cambridge University Press, 1987), and Hal K. Rothman, *The Greening of a Nation* (New York: Harcourt Brace, 1999). For a less scholarly but still academic view, see David Peterson Del Mar, *Environmentalism* (New York: Pearson Longman, 2006), and for a biologist and policy maker's account, see Victor Sheffer, *The Shaping of Environmentalism in America* (Seattle: University of Washington Press, 1991).

Malaria and other diseases began generating their own literatures long before pesticides came on the scene and before scientists learned that some insects transmitted some diseases. Hans Zinnser wrote the classic account of typhus in *Rats, Lice, and History* (Boston: Little, Brown, 1935), approaching the subject from a perspective that would now be called the ecology of disease; that book is worth reading for its point of view, which might usefully be applied to the current debate. The literature on malaria appears under many headings. The most prominent, in addition to "malaria" and "mosquitoes," are tropical disease or medicine, ecology, entomology, and international development. On the malaria-DDT issue, a good place to start is Tina Rosenberg's "What the World Needs Now Is DDT," *New York Times Magazine*, 11 April 2004 and online.

Credits

For permission to reprint material, the author thanks the following organizations and individuals.

"Aerosol Insecticides" and ads from *Soap and Sanitary Chemicals.* Reprinted with permission of *Chemical Week,* which came to include this publication.

Roy J. Barker, "Notes on Some Ecological Effects of DDT Sprayed on Elms." Reprinted with permission of the Wildlife Management Institute.

May Berenbaum, "If Malaria's the Problem, DDT's Not the Only Answer." *Washington Post,* Outlook Section (5 June 2006). Reprinted with permission of the author.

Clarence Cottam and Elmer Higgins, "DDT and Its Effect on Fish and Wildlife." Reprinted with permission of the Entomological Society of America.

Edwin Diamond, "The Myth of the 'Pesticide Menace.'" Reprinted from *The Saturday Evening Post* magazine. © 1963 by the Saturday Evening Post Society. Reprinted with permission.

Paul Dunbar, "The Food and Drug Administration Looks at Insecticides." Reproduced with permission of the Food and Drug Law Institute, Washington, DC, © 1949.

Robert Gillette, "DDT: Its Days Are Numbered, Except Perhaps in Pepper Fields." Reprinted with permission from *Science*, 176 (23 June 1972), 1313–14. © 1972 by AAAS.

Clay Lyle, "Achievements and Possibilities in Pest Eradication." Reprinted with permission of the Entomological Society of America.

Morton Mintz, "'Heroine' of FDA Keeps Bad Drug off Market." © 1962 by *The Washington Post.* Reprinted with permission.

Derek Ratcliffe, "The Status of the Peregrine in Great Britain," *Bird Study*, 10 (June 1963), 56–87, and "Editorial," *Bird Study*, 10 (June 1963), 10. Both reprinted with permission of the British Trust for Ornithology.

Robert Rudd, *Pesticides and the Living Landscape.* Reprinted with permission of the University of Wisconsin Press.

James Stevens Simmons, "How Magic is DDT?" Reprinted from *The Saturday Evening Post* magazine. © 1945 by the Saturday Evening Post Society. Reprinted with permission.

Thomas Sowell, "Intended Consequences." Reprinted with permission of Thomas Sowell and Creators Syndicate, Inc.

Robert S. Strother, "Backfire in the War against Insects." Reprinted with permission from *Reader's Digest*, June 1959. © 1959 by The Readers Digest Assn., Inc.

Robert H. White-Stevens, "Communications Create Understanding." Reprinted with permission of *CropLife.*

Index

WEYERHAEUSER ENVIRONMENTAL BOOKS

The Natural History of Puget Sound Country
by Arthur R. Kruckeberg

Forest Dreams, Forest Nightmares: The Paradox of Old Growth in the Inland West by Nancy Langston

Landscapes of Promise: The Oregon Story, 1800–1940
by William G. Robbins

The Dawn of Conservation Diplomacy: U.S.-Canadian Wildlife Protection Treaties in the Progressive Era by Kurkpatrick Dorsey

Irrigated Eden: The Making of an Agricultural Landscape in the American West by Mark Fiege

Making Salmon: An Environmental History of the Northwest Fisheries Crisis by Joseph E. Taylor III

George Perkins Marsh, Prophet of Conservation by David Lowenthal

Driven Wild: How the Fight against Automobiles Launched the Modern Wilderness Movement by Paul S. Sutter

The Rhine: An Eco-Biography, 1815–2000 by Mark Cioc

Where Land and Water Meet: A Western Landscape Transformed
by Nancy Langston

The Nature of Gold: An Environmental History of the Alaska/Yukon Gold Rush by Kathryn Morse

Faith in Nature: Environmentalism as Religious Quest
by Thomas R. Dunlap

Landscapes of Conflict: The Oregon Story, 1940–2000
by William G. Robbins

The Lost Wolves of Japan by Brett L. Walker

Wilderness Forever: Howard Zahniser and the Path to the Wilderness Act
by Mark Harvey

On the Road Again: Montana's Changing Landscape by William Wyckoff

Public Power, Private Dams: The Hells Canyon High Dam Controversy by Karl Boyd Brooks

Windshield Wilderness: Cars, Roads, and Nature in Washington's National Parks by David Louter

Native Seattle: Histories from the Crossing-Over Place by Coll Thrush

The Country in the City: The Greening of the San Francisco Bay Area by Richard A. Walker

Drawing Lines in the Forest: Creating Wilderness Areas in the Pacific Northwest by Kevin R. Marsh

Plowed Under: Agriculture and Environment in the Palouse by Andrew P. Duffin

Making Mountains: New York City and the Catskills by David Stradling

The Fishermen's Frontier: People and Salmon in Southeast Alaska by David F. Arnold

Shaping the Shoreline: Fisheries and Tourism on the Monterey Coast by Connie Y. Chiang

CYCLE OF FIRE BY STEPHEN J. PYNE

Fire: A Brief History

World Fire: The Culture of Fire on Earth

Vestal Fire: An Environmental History, Told through Fire, of Europe and Europe's Encounter with the World

Fire in America: A Cultural History of Wildland and Rural Fire

Burning Bush: A Fire History of Australia

The Ice: A Journey to Antarctica